GRAVITATIONAL CURVATURE

Brush painting by Jon Frankel

GRAVITATIONAL CURVATURE

An Introduction to Einstein's Theory

Theodore Frankel

UNIVERSITY OF CALIFORNIA, SAN DIEGO

W. H. Freeman and Company
San Francisco

Cover art for softcover addition: *Rimpeling* (*Rippled Surface*)
by M. C. Escher, reproduced with permission from the Escher
Foundation—Haags Gemeentemuseum—The Hague

Sponsoring Editor: Peter L. Renz; Project Editor: Patricia Brewer;
Manuscript Editor: Howard Beckman; Designer: Marie Carluccio;
Production Coordinator: Linda Jupiter; Illustration Coordinator:
Cheryl Nufer; Artist: Boardworks; Compositor: Bi-Comp, Inc.;
Printer and Binder: The Maple-Vail Book Manufacturing Group.

Library of Congress Cataloging in Publication Data

Frankel, Theodore Thomas, 1929–
Gravitational curvature.

Bibliography: p.
Includes index.
1. Relativity (Physics) 2. Gravitation. I. Title.
QC173.55.F7 530.1′1 78-12092
ISBN 0-7167-1006-4
ISBN 0-7167-1062-5 pbk.

AMS(MOS) Subject Classification:
83-01 College level exposition
83-53 Differential geometry

Printed in the United States of America

9 8 7 6 5 4 3 2

to Jonnie

Contents

Preface

In the century since Einstein's birth, Mercury's orbit has precessed a mere forty-three seconds of arc more than could be attributed to Newtonian physics, but the revolution in thought brought about by Einstein's theory (which yielded the correct precession as a first application) has been immense. I try to present, in this book, a way to view the basic aspects of general relativity, a way that is in harmony with my own feelings about geometry and its relationship with pre-quantum physics.

The method of development is "geometric" in the sense that I prefer to compute curvatures or second fundamental forms rather than components of the Riemann tensor or Christoffel symbols. Using this method, simple situations (that is, involving many symmetries) can be handled completely without leaving numerous long and dreary computations "for the reader," as is often done. Differential geometry enters naturally, not only in the very formulation of Einstein's theory but also in the solutions of specific technical problems.

I assume that the reader knows some basic differential geometry. Ideally, the material here would be given as the third quarter of a year course devoted to manifolds and Riemannian geometry. The following is a brief description of the material covered. For the benefit of those already familiar with general relativity, I note, when applicable, how I believe the material differs from most presentations.

Chapter 1 deals with those aspects of special relativity needed for motivational reasons in general relativity and for an understanding of Maxwell's equations.

Chapters 2 and 3 are concerned with the formulation of Einstein's theory and with a heuristic derivation of his equations. Three points are especially emphasized. First, space-time is endowed with a natural metric; this metric can be determined by use of natural clocks ("atomic clocks") and light signals. Second, clock rates are related to the classical Newtonian potential and can be used as a replacement for this potential. Third, Poisson's classical equation $\nabla^2 U = -4\pi\kappa\rho$ can then be translated into the framework of a space-time that is possibly curved: it is found that space-time *must* be curved in the presence of matter. A heuristic "derivation" of Einstein's equations is then given. One noteworthy feature of this derivation is that the divergence-free character of the stress-energy-momentum tensor is a consequence rather than an additional assumption. The material of these chapters is an improvement of a previous effort of mine.

Chapter 4 translates the tensor form of Einstein's equations into several more geometric forms. The resulting equations ("curvature" = "energy density") have been emphasized by J. A. Wheeler many times.

Chapter 5 is concerned with the Schwarzschild solution. The usual procedure is reversed by first deriving the Flamm embedding of the spatial sections.

Chapter 6 reviews background material helpful for the discussion of several topics to be considered. It is assumed here that the reader is familiar with elementary aspects of the exterior differential calculus, as this is clearly the natural tool for several of the topics. A review of interior products and the Lie derivative is presented. The chapter concludes with a brief treatment of the classical motion of a continuum, with emphasis on the Cauchy stress tensor.

Chapter 7 deals with the general relativistic equations of motion and the stress-energy-momentum tensor.

Chapter 8 involves light signals and Fermat's principle in general relativity. A geometric sketch is given of Einstein's expression for the deflection of light by a massive body. The derivation relies on the "Fermat metric" and a preliminary remark about geodesic curvature in conformally related metrics.

Chapter 9 reviews electromagnetism in three-space and in Minkowski space. A distinction is made here between ordinary exterior forms and the "twisted" or "odd" forms introduced by de Rham. A brief review of the Hodge $*$-operator is given. The treatment of Maxwell's equations was inspired by remarks of J. A. Wheeler in that two of Maxwell's equations are taken as axioms and the remaining two are then derived.

Chapter 10 treats electromagnetism in general relativity. One minor point of departure from the usual treatments is the use of the "Fermat metric" for discussing potentials in a static universe.

Chapter 11 investigates Einstein's equations inside matter. A simple derivation of the Schwarzschild interior solution is given. The chapter concludes with a derivation of the Oppenheimer–Volkoff equations that are so important in relativistic discussions of massive spherical bodies and gravitational collapse.

Chapter 12 is devoted to cosmology. The properties of the Friedmann models are derived geometrically. General vorticity-free cosmological models are discussed. The chapter finishes with recent results of Galloway concerning the closure (i.e., compactness) of the spatial part of an expanding universe.

Many topics, e.g., the geometry of the Kruskal extension, the Kerr solution, and the Penrose–Hawking theory of singularities, are not developed in this book. Still, I hope that enough basic material has been covered to allow the reader to go to the literature for further study without too much difficulty. Also, I do not go into technical aspects of physics or astronomy; for examples, I do not mention how the red-shift prediction is "really" tested (via the Mössbauer effect) nor do I discuss how the Hubble parameter is determined.

This book results from my continuing effort to understand the world of relativity. My first glimpse of this world, long ago, was through two superb popular books, Einstein's *Relativity: The Special and General Theory* (1931) and Gamow's *Mr. Tompkins in Wonderland* (1947). In a sense I consider my book an explication, in geometric terms, of material only briefly sketched in the two classics *The Meaning of Relativity* by Einstein (1950) and *Space-Time-Matter* by Weyl (1952).

I am indebted to Professor J. A. Wheeler for helpful comments on an earlier version of this book. I am very grateful to Alan Scales, Professor Richard Matzner, and especially Professor Ann Stehney for their careful reading of the manuscript and their numerous suggestions for improvements and corrections. Even their heroic efforts, however, were perhaps not quite up to the task of coping with my intrinsically careless mathematical style. Thanks are also due to Jill Weaver for her cheerful typing of the manuscript, to Professor Burton Rodin for bringing Irvin's cartoon to my attention, and to Peter Renz and Pat Brewer of W. H. Freeman and Company.

I wish to thank my wife, Jonnie, for the brush painting that forms the frontispiece. My more general feeling of gratitude to her can only be most inadequately conveyed, for, after all, "the dedication that can be expressed is not the perfect dedication."

October 1978
La Jolla

THEODORE FRANKEL

Notation

- Greek indexes run from 1, 2, 3. Roman indexes run from 0, 1, 2, 3.
- Spatial vectors are set in bold type (**A**).
- Space-time vectors and tangent vectors to other manifolds are set in italic type (X).
- Linear transformations and tensors (other than vectors) are set in sans serif type (G). Their components, however, are in italic type.
- Differential forms are usually in Greek (α) or script ($\mathscr{S}$). Note that when dealing with quadratic differential forms, $ds^2 = g_{ij}\, dx^i\, dx^j$, $dx^i\, dx^j \equiv \frac{1}{2}(dx^i \otimes dx^j + dx^j \otimes dx^i)$, while in exterior forms we do not average, $dx^i \wedge dx^j \equiv dx^i \otimes dx^j - dx^j \otimes dx^i$.

The following are specific symbols used in the text. Page numbers refer either to first occurrence or to the definition.

A	action or energy	85
A	magnetic vector potential	125

N unit normal 42
p pressure 31
Q charge 105
r Schwarzschild radial coordinate 47
R scalar curvature 39
R_V scalar curvature of submanifold V 43
$\tilde{R}$ radius of curvature of interior solution 131
$\mathsf{R}(X, Y)$ Riemann curvature transformation 36
R^0_0 component of the Ricci tensor 33
$\mathsf{Ric}\,(X, X)$ Ricci quadratic form 38
S_r sphere of curvature r^{-2} 47
S Cauchy stress tensor 67
$\mathscr{S}$ charge-current 3-form 105
tr trace 43
T stress-energy-momentum tensor 40
T trace of T 32
u unit velocity 4-vector 10
$\tilde{u}$ modified velocity 4-vector 72
U gravitational potential (the sign convention for potentials is opposite to that usually employed by physicists) 22
$u^{\perp}$ subspace of M_p orthogonal to u 71
v classical speed 9
$\mathbf{v}$ classical velocity vector 9

γ Lorentz-Fitzgerald factor 5
Γ Christoffel symbol 36
Δ oriented simplex 103
$|\Delta|$ unoriented simplex 103
$\Delta = 8\pi\kappa\rho - 3h^2$ 148
Δ' cosmological parameter 161
Δ'' cosmological parameter 165
ε_X, ε_i indicators 37
$\epsilon_{i_1 \cdots i_n}$ permutation symbol 106
θ expansion 78
$\theta_{\alpha\beta}$ rate of spatial strain 75
Θ rate of spatial strain 74
$\tilde{\theta}_{ij}$ rate of strain 150
κ Newtonian constant of gravitation 28
κ_i principal curvature 42
λ distinguished or affine parameter 86
Λ cosmological constant (null, in this book) 140
ν coordinate frequency of radiation 21
ν velocity covector 76

$\xi = \sqrt{-g_{00}}$	132
Π projection operator onto $u^{\perp}$	79
$\rho = \mathsf{T}(u, u)$ (curvature) energy density	44
ρ^{*} density	31
ρ_0 rest density	30
σ shear tensor	79
σ charge density	105
σ_F charge density in Fermat metric	126
σ_0 rest charge density	112
τ proper time	9
$\phi = \sqrt{-g_{00}}$	153
ω volume form	37
$\omega^{\perp} = \omega_V$ spatial volume 3-form	77
$\omega_{\alpha\beta}$ spatial vorticity form	75
Ω spatial vorticity form	74
$\tilde{\omega}_{ij}$ vorticity form	150
$\mathbf{A} \cdot \mathbf{B}$ spatial scalar product	8
$\lvert\mathbf{A}\rvert^2 = \mathbf{A} \cdot \mathbf{A}$	
$\langle A, B \rangle$ scalar product	8
$\lVert A \rVert^2 = \langle A, A \rangle$	9
$\lVert A \rVert = [\pm \langle A, A \rangle]^{1/2}$	9
$\lVert X \wedge Y \rVert^2 = \langle X, X \rangle \langle Y, Y \rangle - \langle X, Y \rangle^2$	38
ds^2 (pseudo) Riemannian metric	20
$ds = [\pm ds^2]^{1/2}$	19
$d\tau$ proper time interval	19
$dl_F{}^2$ Fermat spatial metric	92
∇_X covariant derivative	35
$\nabla/\partial\lambda$ absolute or intrinsic derivative	57
$Y^i{}_{;k}$ covariant derivative component	36
∇^2 spatial Laplace operator	28
$\operatorname{grad} U = g^{ij} \dfrac{\partial U}{\partial x^j} \dfrac{\partial}{\partial x^i}$ gradient of U	50
div divergence operator on vectors	59
Div divergence operator on tensors	68
$f_* = \dot{f}$ differential of map f	56
f^* pull back of map f	58
i_X interior product with vector X	58
$[X, Y]$ Lie bracket of vector fields X, Y	57
$\mathscr{L}_X$ Lie derivative	57
exp exponential map for geodesics	72
$*$ Hodge star operator	106
(Σ, N) surface Σ with transverse vector N	110
$\{U\}$ average value of quantity U	66

GRAVITATIONAL CURVATURE

I

Special Relativity

The Lorentz Transformations as Viewed by Einstein

Classical mechanics rests on the notions of absolute space and absolute time. *Absolute space* is assumed to be an affine 3-space with a Euclidean metric that is unique up to constant multiples. One can introduce Cartesian coordinates x, y, z. *Absolute time* is measured by a time coordinate t, unique up to transformations $t \to at + b$, where a and b are constants. Newton's laws of motion are assumed valid when x, y, z, t are used as coordinates for space and time. Newton thought of absolute space as existing independently of matter in the universe, whereas Bishop Berkeley interpreted absolute space as being fixed with respect to the bulk of matter in the universe (the so-called "fixed stars").

An *inertial* (*coordinate*) *system* is a Cartesian coordinate system that moves uniformly, i.e., without acceleration, and, in particular, without rotation, with respect to absolute space. It was realized that Newton's laws of motion are also valid in any inertial system, since accelerations are unaffected by uniform motion. Any two inertial systems are in uniform translational motion with respect to each other.

The existence of an absolute space or absolute frame of reference became dubious much later when highly accurate optical experiments were performed. It was found, toward the end of the last century, that light is propagated *isotropically,* i.e., with the same speed in all directions, in each supposed inertial system. Consider two inertial systems S and S' passing one another. A light pulse is emitted at their common origin at time $t = 0$. It is observed (essentially in the Michelson-Morley experiment) that *both systems see their respective origins as the centers of the resulting spherical light pulse for all time* (the "light pulse paradox")!

This, together with other electromagnetic considerations, led Einstein (among others) to reject the notion of an absolute space. He still retained, however, the notion of a distinguished (but undefined) class of inertial systems. Einstein then showed that this rejection of an absolute space and the resulting notion of absolute motion of an inertial system forces us to abandon also the idea of an absolute time! Einstein (1905) reasoned as follows.

Consider two space-time events E_1 and E_2. When viewed from an inertial system S these events have coordinates (x_1, y_1, z_1, t_1) and (x_2, y_2, z_2, t_2). Since light is propagated isotropically in the system S, the two events will occur simultaneously (at the same time t) in S if and only if light pulses emitted at E_1 and E_2 reach the spatial midpoint $\left(\frac{x_1 + x_2}{2}, \frac{y_1 + y_2}{2}, \frac{z_1 + z_2}{2}\right)$ at the same instant. Consider, for example, two lightning bolts that strike a railway embankment in S at $t = 0$ at spatial coordinates $(x, 0, 0)$ and $(-x, 0, 0)$. These strikes are seen simultaneously at the origin of S at time x/c, where c is the speed of light. Consider an inertial system S' (a train) moving along the tracks (the x direction in S) and suppose that the two ends of the train are struck by the same bolts. If the origin of S' is at the midpoint of the train (equal number of cars forward and behind), it is clear that the forward bolt will be seen at this midpoint before the backward bolt. (It is important here that the speed of light is not infinite.) *Two inertial systems in relative motion will disagree as to whether or not certain spatially separated events are simultaneous.* S and S' *must* be keeping different times. This simple observation by Einstein distinguishes his contribution from those of other workers at the time. (For example, Larmor and Lorentz had already introduced ad hoc "local times" for each inertial system, but this was considered as a purely mathematical convenience.) The light pulse paradox is resolved, for the shape of the pulse is determined by noting the position of the illumination *at a given instant of time!* Einstein's "special theory of relativity," which resulted from this discovery, can be briefly described as follows.

(1) The basic undefined notion is that of the class of distinguished coordinate systems, the inertial systems. Associated with each inertial system is a *proper time* t, unique up to scale changes $t \to at + b$.

(2) Space, when viewed from any inertial system is isotropic. Thus external fields, e.g., electromagnetism, are assumed not to alter the basic structure of space.

(3) Light is propagated isotropically. By a proper choice of the ratios of distance and time scales, all inertial systems can agree that the speed of light is $c \sim 3 \times 10^5$ km/sec. (Note that spatial distances can then be measured by recording the proper time required to send a light beam from the first point to the second and reflecting it back to the first.)

(4) All inertial systems are assumed equivalent for the description of physical laws. No physical experiment will yield an *absolute* motion, and consequently no inertial system is privileged over any other inertial system. (This is the "principle of relativity," first stated in print by Poincaré in 1904.)

Now consider two inertial systems S; x, y, z, t and S'; x', y', z', t'. A particle free from external forces is assumed to execute uniform motion in each inertial system. This insures that any straight line in x, y, z, t coordinates must correspond to a straight line in x', y', z', t' coordinates. It follows that the coordinate functions x', y', z', t' must be *linear* functions of x, y, z, t (we exclude transformations that send finite points to infinity).

A light pulse emitted at a certain event in space-time will be described in S by $dx^2 + dy^2 + dz^2 - c^2\,dt^2 = 0$ and in S' by $dx'^2 + dy'^2 + dz'^2 - c^2\,dt'^2 = 0$. Since the primed coordinates are linear functions of the unprimed ones, these two quadratic expressions in dx, dy, dz, dt must be proportional:

$$dx'^2 + dy'^2 + dz'^2 - c^2\,dt'^2 = k(dx^2 + dy^2 + dz^2 - c^2\,dt^2),$$

where k is a constant (depending on the two inertial systems). Each inertial system has already adjusted the *ratio* of distance and time scales so that the speed of light is c. S' can now make a final distance or time scale choice so that the constant k is unity, i.e.,

$$dx^2 + dy^2 + dz^2 - c^2\,dt^2 = dx'^2 + dy'^2 + dz'^2 - c^2\,dt'^2. \qquad (1\text{-}1)$$

Consider explicitly the following special case. Suppose that S and S' have their x axes parallel and that at the instant $t = 0 = t'$ their origins O and O' coincide. Suppose that S' is moving (with respect to S) in their common x direction with uniform speed v. Isotropy of space and linearity of the transformation demands that the $y'z'$ plane (which is orthogonal to the x' axis in the system S') must also be viewed from S as being orthogonal to the x axis. We may assume then that the y and y' axes and the z and z' axes coincide at $t = 0$ (Figure 1-1). As remarked above, S and S' have already adjusted their ratios (distance scale/time scale) so that the speed of light in each system is c. We can ask further, as above, that they adjust their distance scales so that $y' = y$ (by spatial isotropy they will automatically agree that $z' = z$). We are able to accomplish this because both systems will agree concerning the particular instant at which the y and y' (and z and z') axes overlap. However, the x and x' axes will always overlap; we shall see in a moment that they will not be able to agree on x measurements.

A light pulse emitted at their common origin at $t = 0 = t'$ will satisfy $x^2 + y^2 + z^2 - c^2t^2 = 0$ and $x'^2 + y'^2 + z'^2 - c^2t'^2 = 0$. From $y' = y$ and $z' = z$ we get, from the linear transformation

$$x^2 - c^2t^2 = x'^2 - c^2t'^2$$

with immediate parametric solution

$$\begin{cases} x' = x \cosh u - ct \sinh u \\ ct' = -x \sinh u + ct \cosh u. \end{cases}$$

Since O' (i.e., $x' = 0$) moves with speed v along the x axis, i.e., $x/t = v = c \tanh u$, we have

$$\frac{v}{c} = \tanh u.$$

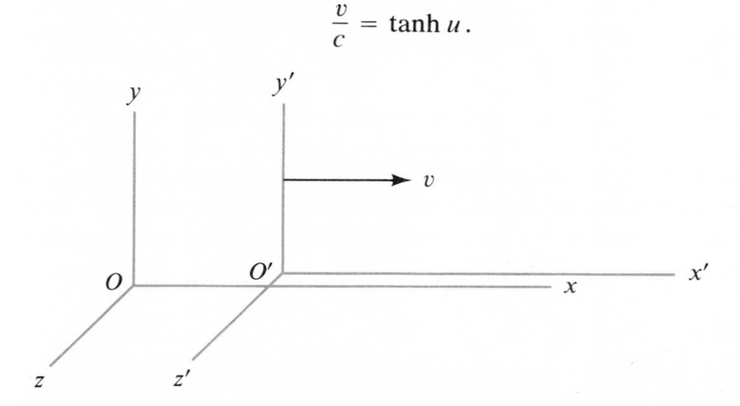

FIGURE 1-1

Finally, putting

$$\begin{cases} \gamma \equiv \cosh u = \left(1 - \dfrac{v^2}{c^2}\right)^{-1/2} \\ \sinh u = \dfrac{v}{c}\gamma, \end{cases}$$

we get the "Lorentz transformation"

$$\begin{cases} x' = \gamma(x - vt) \qquad x = \gamma(x' + vt') \\ t' = \gamma\left(t - \dfrac{vx}{c^2}\right) \qquad t = \gamma\left(t' + \dfrac{vx'}{c^2}\right) \\ y' = y \\ z' = z. \end{cases} \tag{1-2}$$

This transformation is known as a "boost" (in the xt plane). It involves a change of velocity in the x direction but no rotation. The "Lorentz group" is generated by boosts *and* spatial rotations. The group of all transformations satisfying Equation 1-1 is called the Poincaré group. Thus, Poincaré transformations consist of a Lorentz transformation followed by a space-time translation.

The Lorentz transformations of the special type (Eqn. 1-2) can be pictured if the coordinates z and z' are suppressed (Figure 1-2). Here $\tan\alpha = v$, $\tan\beta = v/c^2$, and the "light cone" satisfies $x^2 + y^2 + z^2 - c^2t^2 = 0 = x'^2 + y'^2 + z'^2 - c^2t'^2$. The scales on the axes are not the Euclidean ones, that is, a Euclidean rotation of the xt plane that sends the x axis into the x' axis will not send a unit length in x into a unit length in x'. The scales, rather, are compatible with the relation $x^2 + y^2 + z^2 - c^2t^2 = x'^2 + y'^2 + z'^2 - c^2t'^2$. It is easier to visualize the transformations in Equation 1-2 if the y directions are also suppressed; the resulting diagram is called a Minkowski diagram.

A point in space-time is called an *event*. The history of a (perhaps hypothetical) point particle moving in space-time is a curve of events called the *world line* of the particle.

Given an inertial system S, two events E_1 and E_2 are simultaneous when viewed from S if $t(E_1) = t(E_2)$. It is clear from the Minkowski diagram that they will not be considered as simultaneous when viewed from S' if $v \neq 0$ (Figure 1-3).

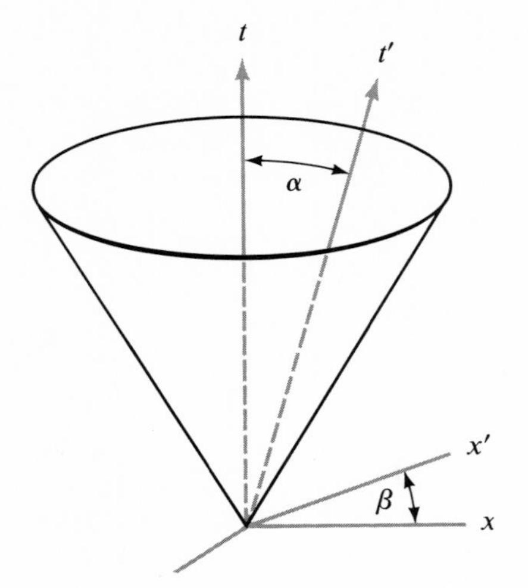

FIGURE 1-2

Consider a measuring rod of length l fixed along the x axis of the S system. The world lines of the ends of the rod are the curves $x = 0$ and $x = l$. To measure the length l' of the same rod from the system S', the position of both ends of the rod must be noted *simultaneously* in the system S', say at $t' = 0$ (Figure 1-4). From Equation 1-2, with $x = l$, $t' = 0$, we see $l = x = \gamma l'$, i.e.,

$$l' = \frac{l}{\gamma} = l\sqrt{1 - v^2/c^2} < l, \tag{1-3}$$

the famous Lorentz-Fitzgerald contraction: "moving" systems see "fixed" lengths contracted in the direction of motion.

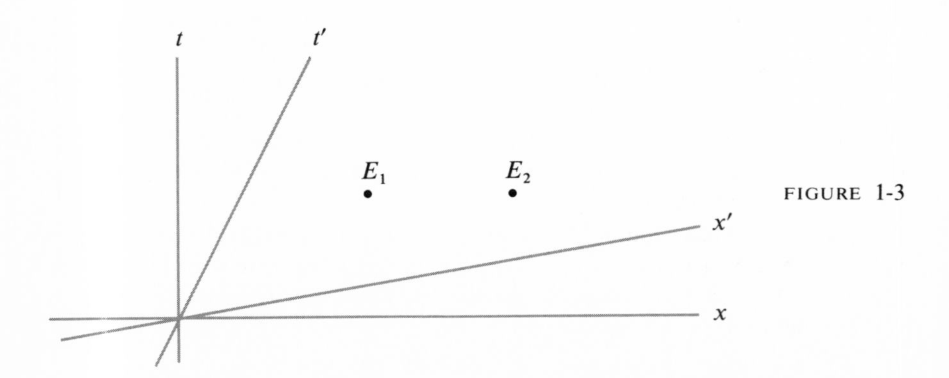

FIGURE 1-3

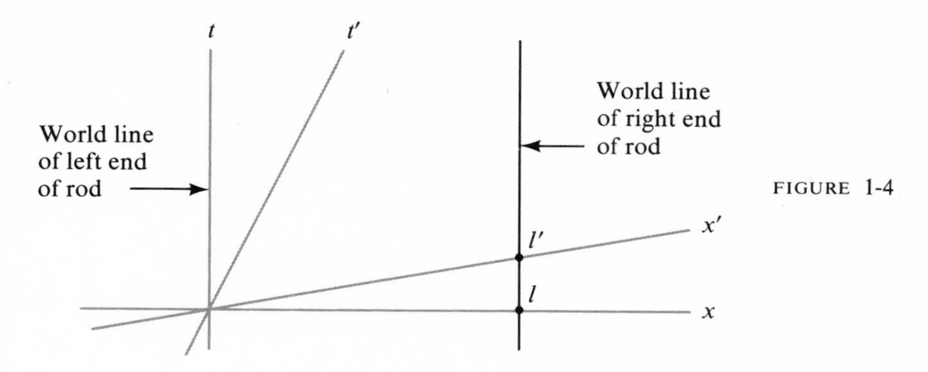

FIGURE 1-4

Consider two events at the same spatial place, the origin of the S system, separated by an S-time of $t = T$. When viewed from the S' system (Figure 1-5), the event with coordinates $x = 0$, $t = T$ has S' coordinates satisfying

$$T' = t' = \gamma t = \frac{T}{\sqrt{1 - v^2/c^2}} > T. \tag{1-4}$$

This is the famous time dilatation of Larmor and Lorentz: "moving" systems see "fixed" clocks as ticking too slowly.

Minkowski Space

In 1908 Minkowski (following earlier work on dynamics by Planck) introduced essentially the following four-dimensional formalism for dealing with dynamics in space and time. (We have already used Minkowski's ideas in the discussion of length contraction and time dilatation.)

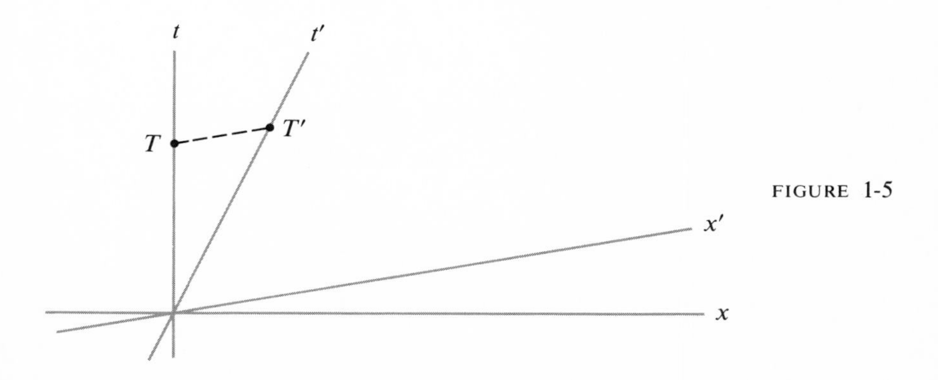

FIGURE 1-5

The space-time manifold M_0^4 of special relativity is an ordinary affine 4-space. However, M_0^4 has a distinguished class of coordinate systems $(t = x^0, x = x^1, y = x^2, z = x^3)$, (t', x', y', z'), ... called *inertial systems*. The coordinate t is called *time*. A point in space-time is again called an *event*. Units can be chosen so that the fundamental constant, the speed of light, is unity, i.e., we shall assume

$$c \equiv 1.$$

M_0^4 is endowed with the *Lorentz* or *Minkowski metric*, which, in any inertial system (x^0, x^1, x^2, x^3) is given by

$$\begin{aligned} ds^2 &= -(dx^0)^2 + (dx^1)^2 + (dx^2)^2 + (dx^3)^2 \\ &= -dt^2 + |d\mathbf{x}|^2. \end{aligned} \tag{1-5}$$

A vector A in M_0^4 (usually called a 4-vector) has a coordinate expression

$$A = A^0 \frac{\partial}{\partial x^0} + A^1 \frac{\partial}{\partial x^1} + A^2 \frac{\partial}{\partial x^2} + A^3 \frac{\partial}{\partial x^3} = \sum_{i=0}^{3} A^i \frac{\partial}{\partial x^i}$$

that is conveniently identified with a column matrix

$$A = \begin{pmatrix} A^0 \\ \mathbf{A} \end{pmatrix} = \begin{pmatrix} A^0 \\ A^1 \\ A^2 \\ A^3 \end{pmatrix}.$$

Here $\mathbf{A}$ is the *spatial part* of A. If B is another vector, their scalar product is defined as

$$\begin{aligned} \langle A, B \rangle &= -A^0B^0 + A^1B^1 + A^2B^2 + A^3B^3 \\ &= -A^0B^0 + \mathbf{A} \cdot \mathbf{B}, \end{aligned} \tag{1-6}$$

$\mathbf{A} \cdot \mathbf{B}$ being the usual scalar product of vectors in R^3. In terms of a "metric tensor" (g_{ij}), $\langle A, B \rangle = \sum_{i,j=0}^{3} g_{ij}A^iB^j$. Thus, in any inertial system

$$(g_{ij}) = \begin{pmatrix} -1 & & & 0 \\ & 1 & & \\ & & 1 & \\ 0 & & & 1 \end{pmatrix}. \tag{1-7}$$

We shall also define $\|A\| = [\pm\langle A, A\rangle]^{1/2}$, the sign being so chosen that $\|A\| \geq 0$. We shall occasionally use the symbol $\|A\|^2$ for $\langle A, A\rangle$.

A parameterized curve $\mathscr{C}$ is of the form $x = x(\lambda)$, that is $x^i = x^i(\lambda)$, $i = 0, 1, 2, 3$, and has tangent vector field $dx/d\lambda$. If $\mathscr{C}$ is the path of a light ray, $ds^2 = 0$ along $\mathscr{C}$, and so $\|dx/d\lambda\|^2 = 0$ for such a path. A vector A for which $\langle A, A\rangle = 0$ is called *light-like*, or *null*. If A has $\langle A, A\rangle < 0$ (resp. $\langle A, A\rangle > 0$) it is called *time-like* (resp. *space-like*). Likewise, for example, a path whose tangent is always time-like is itself called time-like. A time-like path is called a *world line*. If $\mathscr{C}$ is parameterized by t, then the spatial part of the tangent vector $\mathbf{v} = d\mathbf{x}/dt$ is simply the *classical velocity vector*, so that along $\mathscr{C}$ we have

$$\frac{ds^2}{dt^2} = -1 + \left|\frac{d\mathbf{x}}{dt}\right|^2 = -1 + |\mathbf{v}|^2$$

and $|\mathbf{v}| = v$ is the classical speed of the point. The famous relativistic law of mass dependence on speed (see Eqn. 1-10) shows that a material particle with initial speed less than that of light, $v < c = 1$, will always travel at a speed less than that of light. For such a material particle, $ds^2/dt^2 < 0$, that is, $\mathscr{C}$ is always time-like.

Any world line can be reparameterized using its *proper time* parameter τ, defined (up to change of origin) by $ds^2 = -c^2\,d\tau^2 = -d\tau^2$, so that

$$d\tau = \sqrt{-ds^2} = \sqrt{1 - v^2}\,dt. \tag{1-8}$$

Note that if $\mathscr{C}$ is the world line of the origin O' of another inertial system (t', x', y', z'), then $\mathscr{C}$ is described in this new system by $d\mathbf{x}' = 0$, and so by Equation 1-1

$$d\tau = \sqrt{dt^2 - d\mathbf{x}\cdot d\mathbf{x}} = dt',$$

that is, for a particle moving uniformly with respect to an inertial system S, τ is merely the time t' as measured by an inertial system S' moving with the particle! All physical evidence points to the conclusion that τ is the "natural" time coordinate associated with a world line $\mathscr{C}$, whether or not this world line corresponds to a uniform motion with respect to an inertial system. For instance, if $\mathscr{C}: x = x(t)$ is the world line of an unstable particle, then the half-life of the particle refers to time measured by

$$\tau = \int_{\mathscr{C}} \sqrt{1 - v^2}\,dt,$$

not t; moving particles, when viewed from an inertial system, seem to take longer to decay.

If now we use proper time for parameterization of a world line $\mathscr{C}$, we have the special tangent vector $u = dx/d\tau$. In an inertial system S

$$u = \begin{pmatrix} \dfrac{dt}{d\tau} \\ \dfrac{d\mathbf{x}}{d\tau} \end{pmatrix} = \gamma \begin{pmatrix} 1 \\ \mathbf{v} \end{pmatrix}, \tag{1-9}$$

where again $\gamma = (1 - v^2)^{-1/2} = \dfrac{dt}{d\tau}$ · By construction, $\langle u, u \rangle = -1$; u is called the *unit tangent vector* or *velocity 4-vector* to $\mathscr{C}$.

Each particle has associated with it a *rest* or *proper mass* m_0, interpreted as the mass of the particle as seen from an inertial system for which the particle is instantaneously at rest; m_0 is assumed constant. The *momentum* 4-vector for the world line of this particle is by definition $P = m_0 u$. In an inertial system

$$P = m_0 \gamma \begin{pmatrix} 1 \\ \mathbf{v} \end{pmatrix} = \begin{pmatrix} m \\ m\mathbf{v} \end{pmatrix},$$

where

$$m = m_0 \gamma = \frac{m_0}{\sqrt{1 - v^2}} \tag{1-10}$$

is called the *mass* of the particle as seen from the given inertial system. Note that $m \to \infty$ as $v \to c = 1$.

The (*Minkowski*) *4-force* is the vector along the world line defined by

$$f = \frac{dP}{d\tau} = m_0 \frac{du}{d\tau}.$$

In an inertial system

$$f = \begin{pmatrix} \dfrac{dm}{d\tau} \\ \dfrac{d}{d\tau}(m\mathbf{v}) \end{pmatrix} = \begin{pmatrix} \dfrac{dm}{d\tau} \\ \gamma \dfrac{d}{dt}(m\mathbf{v}) \end{pmatrix} = \begin{pmatrix} f^0 \\ \gamma \mathbf{f}_c \end{pmatrix},$$

where $\mathbf{f}_c = d/dt\,(m\mathbf{v})$ is the *classical force* as given by Newton's second law of motion. The term f^0 can be written another way. Since $\langle u, u \rangle = -1$ is constant, the *acceleration 4-vector* $du/d\tau$ is orthogonal to u in the Minkowski metric. $\langle f, u \rangle = 0$ then yields $f^0 = \gamma \mathbf{f}_c \cdot \mathbf{v}$ and

$$f = \gamma \begin{pmatrix} \mathbf{f}_c \cdot \mathbf{v} \\ \mathbf{f}_c \end{pmatrix} = \gamma \left(\mathbf{f}_c \cdot \mathbf{v} \frac{\partial}{\partial t} + \mathbf{f}_c \right) . \tag{1-11}$$

Comparing $f^0 = \gamma \mathbf{f}_c \cdot \mathbf{v}$ with $f^0 = dm/d\tau = \gamma\, dm/dt$ yields $dm/dt = \mathbf{f}_c \cdot \mathbf{v}$, or $dm = \mathbf{f}_c \cdot d\mathbf{x}$.

A first conclusion is, since $m \to \infty$ as $v \to c = 1$, that it takes an infinite classical force to accelerate a particle to the speed of light. Second, the change in mass of a particle is equal to the *classical work* done on the particle, $\int \mathbf{f}_c \cdot d\mathbf{x}$, which classically is the *energy* E imparted to the particle. This and other situations (especially electromagnetism) suggest the *equivalence of mass and energy;* to an amount of energy E of any kind one should associate a mass $m = E$ ($mc^2 = E$ in units where $c \neq 1$). Note that the time component of the momentum 4-vector $m_0 u = \begin{pmatrix} m \\ m\mathbf{v} \end{pmatrix}$ of a particle is then its energy.

The Minkowski Norm

If A is a vector in Minkowski space M_0^4, the quantity $\langle A, A \rangle$ has a remarkable expression in terms of light signals and clocks, an expression that will suggest the metric structure of the actual universe in which we live.

Consider two events, O (the origin of an inertial coordinate system) and E (Figure 1-6). The t axis is the world line of an observer fixed at the spatial origin of the coordinate system. Suppose that a light ray is emitted

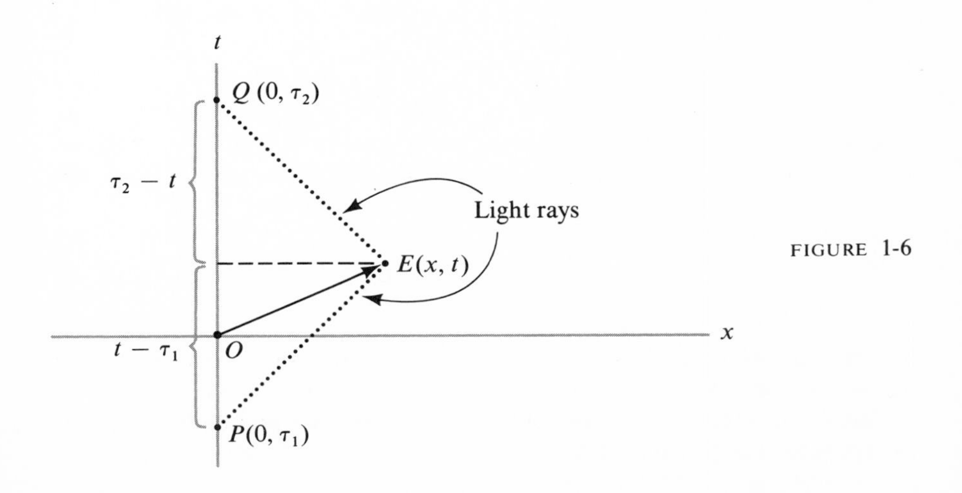

FIGURE 1-6

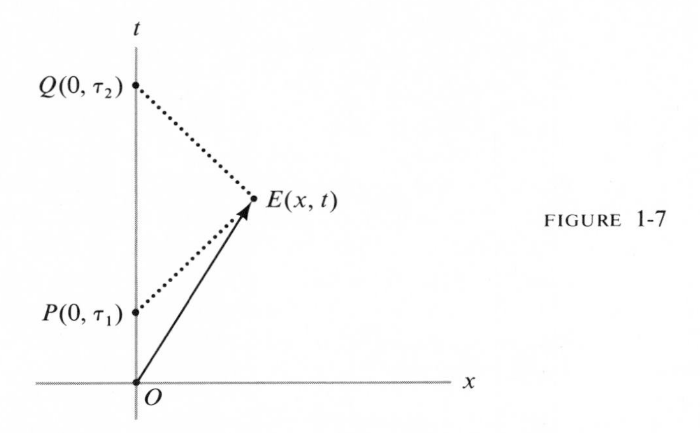

FIGURE 1-7

by this observer at time τ_1 before the event O and that this light ray strikes the event E and is reflected back to the observer, reaching him τ_2 seconds after the event O. Then although τ_1 and τ_2 depend on the inertial system employed, $\tau_1\tau_2$ is independent of the system, and

$$\boxed{\langle \overline{OE}, \overline{OE} \rangle = -\tau_1\tau_2,} \tag{1-12}$$

a formula apparently due to A. A. Robb (1936). For proof, when $\overline{OE}$ is space-like, let E have coordinates (x, t). Then

$$\begin{aligned}\langle \overline{OE}, \overline{OE} \rangle &= x^2 - t^2 = \left[\frac{1}{2}(\tau_2 - \tau_1)\right]^2 - \left[\frac{1}{2}(\tau_1 + \tau_2)\right]^2 \\ &= -\tau_1\tau_2.\end{aligned}$$

Note again, that when E and O occur simultaneously with respect to the inertial system, i.e., when the t coordinate of E is 0, we have for the *spatial distance* between O and E

$$|\overline{OE}| = \|\overline{OE}\| = \tau_1 = \frac{\tau_1 + \tau_2}{2}, \tag{1-13}$$

i.e., the spatial distance between O and E is just half the time required for the inertial observer to receive back a light ray sent to the event E.

The proof of Robb's formula when $\overline{OE}$ is time-like proceeds as above, except that τ_1 is positive (Figure 1-7).

2

Clocks and Gravitational Potential

Gravitation, Acceleration, and the Principle of Equivalence

Compare electromagnetism and gravitation. Space seems to exist independently of any electromagnetic fields present. A charged particle responds to an electromagnetic field, but an uncharged particle moves in space as if the field were not present at all. Not so with gravitation! *All* particles react to the gravitational field, and in fact they all react in the same manner, independent of their mass or composition. Near the earth's surface this is exemplified by Galileo's law of falling bodies. The classical experiments of Eötvös, and later improved versions, verify with remarkable accuracy that the gravitational acceleration of a body is indeed independent of its mass.

Newton's gravitational theory is a *scalar* theory; the gravitational field is described by a single function, the gravitational potential function. If a mass vibrates "here," its effect is felt everywhere *instantaneously*.

However, there has been strong feeling for a long time that this "action at a distance" is unsatisfactory. Especially since the advent of special relativity, it was felt that no signal could ever be propagated faster than the speed of light, for such a signal would violate "causality." Consider such a signal sent from the origin O of an inertial system S to an event E (Figure 2-1). Then by a Lorentz transformation one can find an inertial system S' for which $t'(E) < 0$. Thus, the system S' would see the signal arrive at E before it is sent from O!

Now, while a *static* electric field is governed by a single scalar "Coulomb potential," nonstatic electromagnetic fields require a scalar potential and a vector potential, or, briefly, a 4-vector potential (this is discussed in Chapter 10, p. 121ff). A consequence of this is that electromagnetic disturbances are propagated with the speed of light. (Briefly, the potentials were shown by Maxwell to satisfy a wave equation rather than a potential or Poisson equation; see Chapter 10, p. 127.) This suggests that perhaps gravitation also should be described by a vector potential or perhaps a still more complicated "tensor" potential. This view was supported by astronomical evidence predating relativity, namely, the precession of the perihelion of Mercury's orbit, that indicated that Newton's law of gravitation is indeed incorrect when dealing with moving bodies, though the error is exceedingly small.

The notion of an inertial system, so crucial for both classical mechanics and Minkowski space, is undefined. Attempts to exhibit such systems in the real world have all failed. One is perhaps tempted to say that an inertial system is one in which Newton's laws of motion hold, but this is circular since the definition of force in Newton's laws $\mathbf{f} = d/dt(m\mathbf{v})$ clearly depends on the system of coordinates used. If Newton's laws

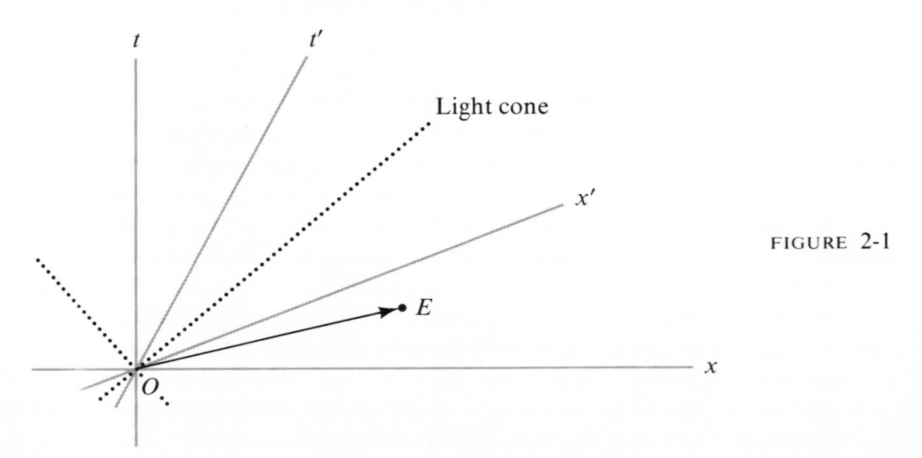

FIGURE 2-1

are "satisfied" in a particular system, and if we pass to a new system accelerating with respect to the first, Newton's laws will hold in this new system only if so-called fictitious forces are included (if rotation is involved these are the centrifugal and Coriolis forces). It is not clear what a "fictitious" force is except that it must have the property that the "resulting" acceleration must be independent of the mass or composition of the body.

A collection of particles in a stationary elevator near the earth's surface is observed to accelerate downward due to the force of gravity, but if the elevator is allowed to fall freely the particles seem (to great approximation) not to accelerate at all with respect to the elevator. This is simply an expression of Galileo's law (acceleration is independent of mass). Is gravitation itself, then, a fictitious force, appearing merely because of a poor choice of coordinate system? Is a "freely falling" coordinate system an inertial system? This cannot be exactly true, for even neglecting the small gravitational attraction of the particles, they will seem to approach each other very slightly as they fall toward the earth's center, i.e., because of the small inhomogeneity of the earth's gravitational field. Also, two such systems will not have their coordinates related by *linear* transformations, due to their acceleration. We apparently must give up the notion of any preferred coordinate system. Einstein strongly believed that the laws of physics must be stated in a general form valid in *any* coordinate system, not just the inertial systems of special relativity, hence the term "general relativity."

Galileo's law made a profound impact on Einstein. In Newton's second law of motion $\mathbf{f} = m_I\mathbf{a}$, the *inertial mass* m_I of the test body is a measure of that body's inertia or resistance to acceleration. On the other hand, if $\mathbf{f}$ is due to the gravitational attraction of other matter, then $\mathbf{f} = m_G\mathbf{k}$. Here $\mathbf{k}$ depends only on the other matter, and the *gravitational mass* m_G is a measure of the test body's gravitational interaction with other bodies. From these we have $\mathbf{a} = (m_G/m_I)\mathbf{k}$ and Galileo's law then implies that m_G/m_I is in fact independent of the composition of the test body. This proportionality of gravitational and inertial mass could not in Einstein's view be a coincidence. Note, for instance, that in the electrical analogy m_I is again the mass of the test body while m_G is replaced by the electric charge e, and certainly e/m depends very much on the composition of the test body. Einstein felt that an acceptable theory of gravitation must have Galileo's law as a natural consequence, and we shall see in Chapter 7 how general relativity accomplishes this.

The "natural" state of a particle seems to be free fall; in free fall it feels no "force of gravity." It is only when its free fall is interrupted that the "fictitious" force of gravity is felt. Acceleration should connote a motion that differs from free fall. Einstein noted that locally the effects of acceleration with respect to an inertial system are approximately the same, or "equivalent," to those of gravitation in an inertial system of special relativity, and one can use this *principle of equivalence* heuristically to predict gravitational effects merely by observing effects due to acceleration (see e.g., Einstein, 1931). Again we are led to consider gravitation as an intrinsic aspect of space-time itself, and not some field that exists in some Minkowski space. Minkowski space is but a local idealization of the true space-time. What then *is* the structure of space-time?

The Pseudo-Riemannian Structure of Space-Time

Let us assume that space-time is a four-dimensional manifold, M^4, the points of which are again called *events*.

Usually it is hypothesized that M^4 carries a "pseudo-Riemannian" metric, i.e., a generalized scalar product in each tangent space that assumes the form of Equation 1-6 for an appropriate basis of the tangent space. We prefer to propose a specific pseudo-Riemannian metric that will lead, via heuristic reasoning, to a "derivation" of Einstein's basic results. After the derivation of these field equations, the specific manner in which the metric is chosen will play no essential role in further developments. For the construction of the metric we proceed as follows (see Ehlers, Pirani, and Schild, 1972, for a related development).

At each $p \in M^4$ consider the collection of all segments of light ray paths through p that lie in a small neighborhood of p. The tangents to these paths at p form the *light cone* in the tangent space M^4_p to M^4 at p. We assume that the complement of this cone in M^4_p consists of two disjoint regions, the *inside* and *outside* of the light cone (Figure 2-2). We further require that the inside consist of two disjoint connected regions, the future and past parts. (In R^4, this latter requirement is satisfied by the "normal" cone $-t^2 + x^2 + y^2 + z^2 = 0$, but not by the cone $-t^2 - x^2 + y^2 + z^2 = 0$.) This cone structure is to vary smoothly over the manifold, though future and past parts may be interchanged upon going around loops in the manifold. If future and past parts are maintained around all loops, M^4 is said to be *time orientable*.

Vectors at p pointing to the inside of the light cone at p are called time-like; those pointing outside are space-like. Vectors tangent to the cone at p are light-like, or null. A world line is a curve in M^4 whose tangent is always time-like.

Imagine an "observer" associated with each world line who can send out light signals. We assume the observer carries an *atomic clock,* by

"People slowly accustomed themselves to the idea that the physical states of space itself were the final physical reality."

—Professor Albert Einstein

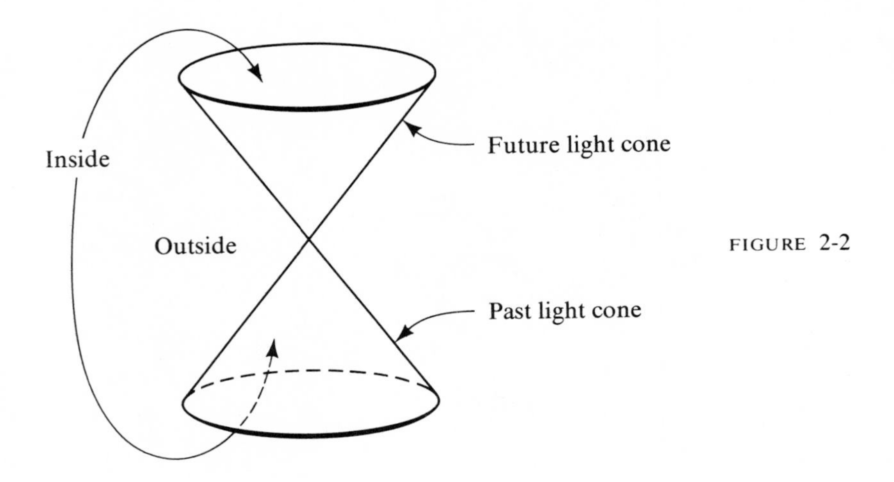

FIGURE 2-2

which we mean the following. Take an atom of specified type, e.g., hydrogen, moving along the given world line. An electron of this atom in a specified excited state will, when falling to a specified lower energy state, emit radiation of a certain definite frequency. This frequency is used to define time intervals, i.e., it can be used as a clock. For this definition to be useful it is crucial that the frequency be independent of the past history of the atom. Atoms at different points may emit different frequencies when viewed from some point, but if the atoms are brought together and made to travel along the same world line their frequencies are observed to coincide. An atomic clock on the earth's surface yields time intervals that, to a high degree of accuracy, are proportional to ordinary (astronomical) time. However, if we have two atoms at the same point, one at rest and one in motion with respect to an observer, the observer will detect different frequencies that satisfy a Doppler effect.

The time τ kept by an atomic clock along a world line $\mathscr{C}$ is called the *proper time* parameter for $\mathscr{C}$.

Our proposal for a pseudo-Riemannian metric for M^4 continues as follows. Let v be a vector, i.e., a tangent vector to M^4, at the event O. Let $E\colon (-1, 1) \to M^4$; $\varepsilon \to E(\varepsilon)$ be a curve thru $O = E(0)$ having v as tangent at O, $\dot{E}(0) = v$. Let $\mathscr{C}$ be the world line of any observer passing thru O, $\mathscr{C}$ being parameterized by proper time τ, that is, by an atomic clock, and having $\mathscr{C}(0) = O$. By the same process used in Minkowski space (p. 11), send a light ray from $\mathscr{C}$ to the event $E(\varepsilon)$ and then back to $\mathscr{C}$ (Figure 2-3). We shall assume that for small ε, the emission and reception events are unique. Form $\tau_1(\varepsilon)\ \tau_2(\varepsilon)$ for each small ε. Let us hypothesize that the "norm"

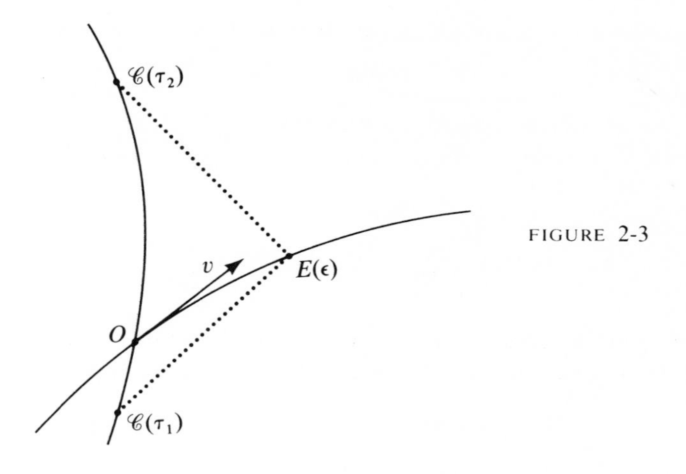

FIGURE 2-3

$$\langle v, v \rangle \equiv \lim_{\varepsilon \to 0} \frac{-\tau_1(\varepsilon)\tau_2(\varepsilon)}{\varepsilon^2}$$

exists, independent of the choices made of the curve E and the world line $\mathscr{C}$. It can be shown (Ehlers, Pirani, and Schild, 1972; Galloway, 1976) that certain mathematical assumptions together with the "weak principle of equivalence" (Galileo's law of uniqueness of free fall) show that this "norm" is indeed the norm induced by a pseudo-Riemannian metric ds^2, i.e., $\langle v, v \rangle$ has the form of Equation 1-6 for a suitable basis of the tangent space M_0^4. ***This is the metric we shall use for M^4. Light signals and clocks are all that we need for its determination!*** To determine whether M^4 is flat Minkowski space or some intrinsically curved space-time, we need to interpret our metric in terms of classical Newtonian gravitation.

If $\mathscr{C} = \mathscr{C}(\lambda)$ is a space-like or time-like curve in M^4, we define

$$ds = \left[\pm \left\langle \frac{d\mathscr{C}}{d\lambda}, \frac{d\mathscr{C}}{d\lambda} \right\rangle\right]^{1/2} d\lambda$$

along $\mathscr{C}$, where the negative sign is used if $\mathscr{C}$ is time-like. The integral $\int_{\mathscr{C}} ds$ is independent of the parameter λ of $\mathscr{C}$. When $\mathscr{C}$ is space-like, $\int ds$ defines *arc length* along $\mathscr{C}$. When $\mathscr{C}$ is time-like we write $d\tau = ds$, so that $(d\tau)^2 = -ds^2$. Note that from the metric construction given above, $\int_{\mathscr{C}} ds = \int_{\mathscr{C}} d\tau$ is merely the proper time kept by an atomic clock having $\mathscr{C}$ as its world line! (In our construction choose $\lambda = \varepsilon$ to be proper time along

$\mathscr{C}$ and choose $E(\varepsilon) = \mathscr{C}(\varepsilon)$; note then that $\tau_2(\varepsilon) = \varepsilon = \tau_1(\varepsilon)$, and so $\left\langle \frac{d\mathscr{C}}{d\lambda}, \frac{d\mathscr{C}}{d\lambda} \right\rangle = -1$.)

In terms of arbitrary local coordinates $t = x^0, x^1, x^2, x^3$ in M^4 (subject only to $\partial/\partial x^0$ time-like and $\partial/\partial x^1$, $\partial/\partial x^2$, $\partial/\partial x^3$ space-like), the metric has a local expression

$$ds^2 = g_{00}(x)\, dt^2 + 2 \sum_{\beta=1}^{3} g_{0\beta}(x)\, dt\, dx^\beta + \sum_{\alpha,\beta=1}^{3} g_{\alpha\beta}(x)\, dx^\alpha\, dx^\beta.$$

If an observer is "spatially fixed" with respect to this coordinate system, i.e., x^1, x^2, x^3 are constant along his world line, we have $ds^2 = g_{00}\, dt^2$, and so

$$d\tau = \sqrt{-g_{00}}\, dt$$

tells us how proper time τ (as kept by the observer's atomic clock) along the world line is related to the rather arbitrary "coordinate time" t.

Gravitational Potential

What relation exists between the metric coefficients g_{ij} and the distribution of matter and energy in the universe? To get some idea of the connection we need only consider a simple physical situation. An especially simple geometric situation occurs when the local coordinates x^i, $i = 0, 1, 2, 3$, can be chosen such that the metric coefficients g_{ij} are independent of $t = x^0$. In this situation the metric ds^2 is invariant under time translations $t \to t + h$, i.e., time translations are *isometries,* and the vector field $u = \partial/\partial t$ generates a one-parameter group of isometries; u is called a *Killing vector field* (after the geometer Killing). A universe so described is called a *stationary* universe. If, furthermore, the cross terms $g_{0\beta}$ vanish, $\beta = 1, 2, 3$, we say the universe is *static*. The vanishing of the cross terms $g_{0\beta}$ implies that the *spatial sections* V_{t_0}, obtained by putting $t = \text{constant} = t_0$, are orthogonal to the *time lines,* obtained by putting $x^1 = \text{constant}$, $x^2 = \text{constant}$, and $x^3 = \text{constant}$, i.e., the Killing vectors $u = \partial/\partial t$ are orthogonal to the spatial sections. For a static universe we then have

$$ds^2 = g_{00}(x^1, x^2, x^3)\, dt^2 + \sum_{\alpha,\beta=1}^{3} g_{\alpha\beta}(x^1, x^2, x^3)\, dx^\alpha\, dx^\beta.$$

We envision this universe as having all matter and energy (except for idealized infinitesimal test particles, photons, etc.) at rest with respect to the given coordinate system.

Stationary and static universes can be described in an equivalent fashion, as follows. Let an M^4 admit a nonvanishing time-like Killing vector field X. One can always introduce coordinates $t = x^0, x^1, x^2, x^3$ locally such that $X = \partial/\partial t$ and the components g_{ij} are independent of t, and so M^4 is stationary. If, further, the distribution $X^\perp$ of 3-planes orthogonal to X is *integrable,* then M^4 is static. Each integral curve $\mathscr{C}$ of the Killing vector field $X = \partial/\partial t$ may be considered as the world line of an observer. The 3-plane $X_t^\perp$ at $\mathscr{C}(t)$ can be described, as in special relativity, as *the* spatial 3-plane at the event $\mathscr{C}(t)$ for this observer; curves tangent to $X_t^\perp$ are curves of events that are "infinitesimally simultaneous" with $\mathscr{C}(t)$. Since $X = \partial/\partial t$ generates isometries and $\partial/\partial t$ is invariant under these isometries, the 3-planes $X^\perp$ are also invariant under the isometries. As we shall see in Chapter 7, the assignment $t \to X_t^\perp$ yields, in a well-defined fashion, a one-parameter group of *rotations* of an associated 3-space. We can then say that in a stationary universe world lines "infinitesimally close" to a given world line are "spatially rotating" about this world line. When the rate of rotation is zero, the universe is static.

Consider then a static universe corresponding to a static bounded mass distribution, that is, assume that $M^4 = R \times V^3$, where V^3 is topologically R^3 and where V^3 contains a static bounded mass distribution. Let us assume that the metric ds^2 tends to the Minkowski metric as $|\mathbf{x}| \to \infty$

$$ds_\infty^2 = -dt^2 + |d\mathbf{x}|^2.$$

Suppose that an atom at spatial coordinates $\mathbf{x}$ emits radiation with coordinate frequency ν, i.e., ν wave crests per *coordinate* second are emitted. Because translation in time is an isometry, it follows that the wave crests arrive at any other point again with *coordinate* frequency ν (Figure 2-4). (In this case we must assume that the emitted radiation has a negligible effect on the metric of space-time. Then if $\mathscr{C}$ is a space-time path of constant phase, e.g., a wave crest, then translating $\mathscr{C}$ in time by an amount $1/\nu$ will yield automatically the next wave crest path.) Now at the emission point $\mathbf{x}$ we have (since $\mathbf{x}$ is spatially fixed) $d\tau = \sqrt{-g_{00}(\mathbf{x})}\, dt$, and so the *proper* frequency of the radiation (number of wave crests emitted per proper second) is $\nu/\sqrt{-g_{00}(\mathbf{x})}$. Thus the energy of the radiation, by the Einstein-Planck law, is

$$\varepsilon(\mathbf{x}) = \frac{h\nu}{\sqrt{-g_{00}(\mathbf{x})}} \cdot$$

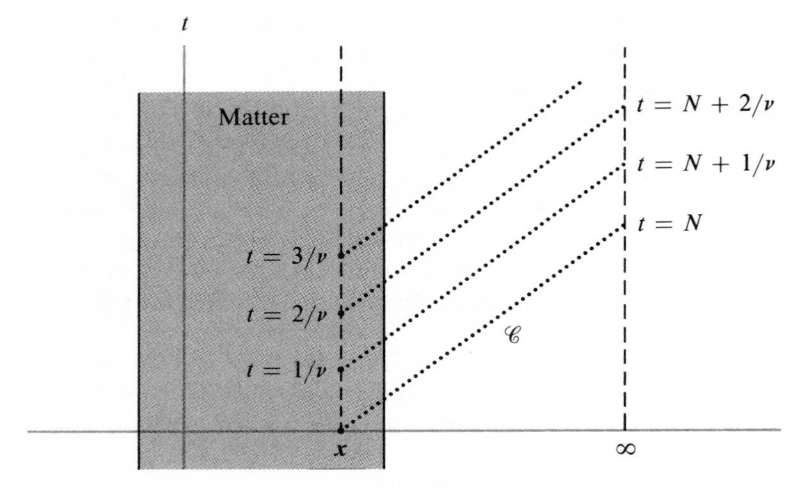

FIGURE 2-4

However, the energy received "at infinity," i.e., at a great distance from **x**, is given by

$$\varepsilon(\infty) = \frac{h\nu}{\sqrt{-g_{00}(\infty)}} = h\nu.$$

The "gravitational potential" at **x** should be the work done in moving a unit mass from **x** to infinity (note that physicists usually call the negative of this quantity the potential). Equivalently, this potential $U(\mathbf{x})$ should be the energy lost by the radiation in going from **x** to infinity, divided by the energy of the radiation. It is not clear whether we should divide by the energy at **x** or at infinity. With hindsight we adopt the former:

$$U(\mathbf{x}) = \frac{\varepsilon(\mathbf{x}) - \varepsilon(\infty)}{\varepsilon(\mathbf{x})} = 1 - \sqrt{-g_{00}(\mathbf{x})}. \tag{2-1}$$

In a static universe with metric approaching the Minkowski metric at infinity, we are led to consider $1 - \sqrt{-g_{00}(\mathbf{x})}$ *as the appropriate analog of the classical Newtonian gravitational potential.*

We note that $U(\mathbf{x})$ can be interpreted in terms of the "speed of light." Recall that we have chosen units such that the speed of light, using *proper* time, is unity. That is, if p and q are "infinitesimally near" points in the same spatial section V_t^3, then the spatial distance between them is just half the *proper* time required for a light ray to go from one to the other and back again (see Chapter 1 for the case of Minkowski space). The speed of light does *not* turn out to be one if we use the *coordinate* time t. Since the path of a light ray satisfies $ds^2 = 0$, we have in our static universe $g_{00}(\mathbf{x})\, dt^2 + \Sigma g_{\alpha\beta}(\mathbf{x})\, dx^\alpha\, dx^\beta = 0$. Since

$$dl^2 \equiv \sum_{\alpha,\beta=1}^{3} g_{\alpha\beta}(\mathbf{x})\, dx^\alpha\, dx^\beta \tag{2-2}$$

is the induced Riemannian metric in each spatial section t = constant, the light ray path satisfies

$$\frac{dl}{dt} = \sqrt{-g_{00}}, \tag{2-3}$$

indicating that the *coordinate* speed of light is merely

$$c(\mathbf{x}) = \sqrt{-g_{00}(\mathbf{x})},$$

so that

$$U(\mathbf{x}) = 1 - c(\mathbf{x}).$$

It is remarkable that Einstein in 1912, following earlier work of 1907 and three years before his final formulation, proposed $c(\mathbf{x})$ as a replacement for the Newtonian gravitational potential in the classical Poisson equation. This proposal was based on his newly discovered "principle of equivalence" (see p. 16, top). Einstein soon abandoned this approach; he was not familiar with Riemannian geometry at this time and consequently his analysis took place in what would now be considered as a space-time with metric $ds^2 = -c^2(\mathbf{x})\, dt^2 + dx^2 + dy^2 + dz^2$ having *flat* spatial sections. We shall see, in the next chapter, that this approach can succeed if one is willing to allow *curvature* of the metric. It is also fair to say that the specific appropriate geometric results needed were not available in 1912. The real key to the connection between potential and curvature was supplied by Levi-Civita only *after* Einstein had given the correct solution to the problem of gravitation. Einstein did not "derive" his solution in the sense of a direct link between theories of the past and new axioms; rather, he "presented" it, based on his magnificent insight into the nature of physical laws. His solution can be better understood in terms of the above mentioned connection between potential and curvature, by inverting Levi-Civita's arguments and combining them with results that were understood only afterward. (For the turbulent history of the struggle to discover the correct field equations, see Mehra, 1974.)

Let us return to our analysis. If we recall that a radiating atom is assumed to keep proper time, we can look at Equation 2-1 in the following manner. A certain atom in a certain excited state at $\mathbf{x}$ radiates with coordinate frequency $\nu_{\mathbf{x}}$ or proper frequency $\nu_{\mathbf{x}}/\sqrt{-g_{00}(\mathbf{x})}$. This radiation is received at another point $\mathbf{y}$ again with coordinate frequency $\nu_{\mathbf{x}}$. The same type of atom in the same excited state at $\mathbf{y}$ will radiate, by definition of proper time, with the same proper frequency $\nu_{\mathbf{x}}/\sqrt{-g_{00}(\mathbf{x})}$, i.e., with coordinate frequency

$$\begin{aligned} \nu_{\mathbf{y}} &= \nu_{\mathbf{x}} \frac{\sqrt{-g_{00}(\mathbf{y})}}{\sqrt{-g_{00}(\mathbf{x})}} \\ &= \nu_{\mathbf{x}} \frac{1 - U(\mathbf{y})}{1 - U(\mathbf{x})} \,. \end{aligned} \tag{2-4}$$

This is Einstein's *red shift formula*—atoms radiate more slowly in a strong gravitational field. Einstein first attempted to apply this to the gravitational field of the sun, using for U the Newtonian potential $\kappa M_{\odot}/r$, where κ is the gravitational constant and r is the distance from the sun's center. The effect leads to a shift in the spectral lines toward the red.

It is interesting that an "equivalence principle" argument first used by Einstein can be modified to yield again that $U = 1 - \sqrt{-g_{00}}$ corresponds to the Newtonian potential. Consider a "freely falling" system S at so great a distance from the bulk of gravitating matter that it can be considered an approximation to an inertial system of special relativity. There are fixed observers in this system, all keeping the same inertial time t. Now let us consider another system S' consisting of a disc that is rotating with uniform speed ω about the z axis of the S system. Observers on this disc will suffer a centrifugal acceleration $\omega^2 r$, where r is the distance from the center of the disc. If they do not believe in absolute motion they could attribute this force to an equivalent (but strange) gravitational force that is pulling them out to the rim. From special relativity we know that their proper clocks run at a rate $d\tau = \sqrt{1 - v^2}\, dt = \sqrt{1 - \omega^2 r^2}\, dt$, which we write as $d\tau = \sqrt{-g_{00}(r)}\, dt$. The clock at the center of the disc runs at the same rate as the inertial clocks, i.e., $g_{00}(O) = -1$. Suppose now that a person at the center of the disc releases a body attached to a string. The body is released at a very slow rate and is allowed to slide down a smooth trough to an observer at radial distance R (Figure 2-5). The centrifugal acceleration $\omega^2 r$ together with the mass m of the body give rise to a force $m\omega^2 r$ in the string that is felt by the central observer. Again, from special relativity (Eqn. 1-10), if m_0 is the mass of the body at the center, then its

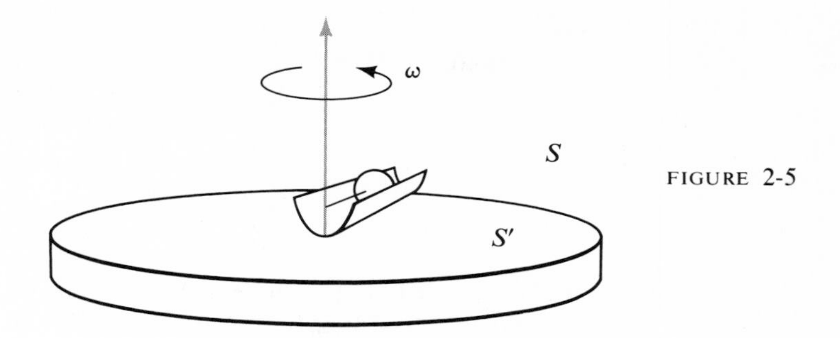

FIGURE 2-5

mass at radial distance r is $m_0(1 - v^2)^{-1/2} = m(r)$. The work done by the central observer in letting the body slowly out to $r = R$ is

$$W = \int \text{force } dr = \int_0^R \frac{m_0\omega^2 r\, dr}{\sqrt{1 - \omega^2 r^2}}$$

$$= m_0(1 - \sqrt{1 - \omega^2 R^2}) = m_0(1 - \sqrt{-g_{00}(R)}).$$

The work done per unit mass, W/m_0, is then the potential $U(R)$, as desired. Since the center of the disc corresponds to the point at infinity in this inverted "gravitational field," we again conclude that the "Newtonian potential" is measured by the difference in rates of atomic clocks, one at infinity, the other at the point in question.

Einstein (1931, p. 157) connected atomic clock rates with potential but without using the special relativistic expression for the variation of mass. Rather than using $U = 1 - \sqrt{-g_{00}(R)}$ for his potential, Einstein used

$$V = \int_0^R \omega^2 r\, dr = \frac{\omega^2 R^2}{2} = \frac{1}{2} + \frac{1}{2} g_{00}(R).$$

Since g_{00} is ordinarily very close to the Minkowski value -1 (as we shall see later), these two expressions will differ by an exceedingly small amount. For our purposes, however, the square root will play an important role.

Is Gravitation Governed by a Single Potential?

It may seem far-fetched to draw conclusions about gravitation from a situation involving centrifugal force, but it was noted long ago (by

Bishop Berkeley and especially Ernst Mach) that *centrifugal force has a gravitational origin!* If there existed only the systems S and S' and no other matter in the universe, then it would be highly unlikely that centrifugal force could be felt in S' and not S; who is to say who is rotating and who is at rest? It is only because there *is* distant matter in the universe that we can distinguish between a ''freely falling'' system S and a system S' whose molecular forces are resisting the gravitational field of the distant matter.

Consider an idealized universe consisting of the earth surrounded at a great distance by a large uniform mass distribution, the ''fixed stars,'' revolving once about the earth each 24 hours. Consider a Foucault pendulum at the North Pole. If the pendulum is not swinging the bob reacts essentially only to the gravitational pull of the earth; the effect of the distant matter is negligible. On the other hand, if the pendulum is swinging the bob does react to the distant matter, keeping its plane of oscillation approximately fixed with respect to this matter. This situation is reminiscent of the situation in electromagnetism in which a charged particle ''at rest'' in a stationary electromagnetic field responds only to the electric field vector $\mathbf{E} = \text{grad}\ \varphi$, while a charge ''in motion'' also responds to the magnetic field $\mathbf{B} = \text{curl}\ \mathsf{A}$ (see Chapter 9, p. 100). The metric coefficient $\sqrt{-g_{00}}$ seems to play a role in gravitation similar to that of the electrostatic potential φ. It could be reasoned that the other coefficients (g_{ij}) would play the role of potentials for other effects (e.g., the Coriolis ''force'' involved in the Foucault pendulum) similar to that of the vector potential $\mathbf{A}$ of magnetism. Einstein's quantitative solution to this problem, the most remarkable intellectual achievement of all time in the opinion of this author, is discussed in the next chapter.

3

A Heuristic Derivation of Einstein's Equations

Poisson's Equation

Consider a universe M^4 in which the only matter or energy present consists of a blob of fluid. Gravitation will pull the molecules of the fluid together and pressure will prevent the blob from collapsing to a single point. Assume the equilibrium state to be a spherical blob at rest. This seems to be the simplest physically realistic model to consider.

After equilibrium takes place there is no "motion"; we therefore assume that the metric of space-time M^4 is static:

$$ds^2 = g_{00}(x^1, x^2, x^3)\, dt^2 + dl^2,$$

where $dl^2 = g_{\alpha\beta}(x^1, x^2, x^3)\, dx^\alpha\, dx^\beta$ is the metric for the spatial sections V_t^3 resulting from putting $x^0 = t =$ constant. Here we use the "Einstein summation convention": a repeated index, once covariant and once contravariant in any expression is to be summed. We also use the convention that Greek indexes are taken from 1, 2, 3, while Latin indexes are from 0, 1, 2, 3 whenever we consider a space-time manifold M^4.

It should be emphasized that we are not implying that space-time is curved; we are simply not insisting that it be flat! We assume that the spherical ball is centered at the origin of coordinates and it is natural to use spherical coordinates r, θ, φ in the spatial sections V^3 (we do *not* interpret r as distance from the origin; distances are measured by the yet to be determined metric dl^2). By spherical symmetry

$$ds^2 = g_{00}(r)\, dt^2 + g_{\alpha\beta}(r, \theta, \varphi)\, dx^\alpha\, dx^\beta,$$

where $x^0 = t$, $x^1 = r$, $x^2 = \theta$, $x^3 = \varphi$.

In the previous chapter we gave heuristic reasons for believing that $U = 1 - \sqrt{-g_{00}(r)}$ should play the role of Newtonian potential in the spatial sections V^3. In Euclidean space the Newtonian potential satisfies Poisson's equation

$$\nabla^2 U = -4\pi\kappa\rho^*, \tag{3-1}$$

where κ is the gravitational constant and ρ^* is the density of mass. Now in any Riemannian manifold one defines the Laplacian of a function f by either of the equivalent expressions

$$\begin{aligned}\nabla^2 f &= g^{\alpha\beta} f_{;\alpha\beta} \\ &= \frac{1}{\sqrt{g}} \frac{\partial}{\partial x^\alpha} \left(\sqrt{g}\, g^{\alpha\beta} \frac{\partial f}{\partial x^\beta} \right).\end{aligned}$$

In the first expression $f_{;\alpha} = \dfrac{\partial f}{\partial x^\alpha}$ and $f_{;\alpha\beta}$ is the βth covariant derivative of the covariant vector $f_{;\alpha}$, that is,

$$f_{;\alpha\beta} = \frac{\partial}{\partial x^\beta} \left(\frac{\partial f}{\partial x^\alpha} \right) - \frac{\partial f}{\partial x^\gamma} \Gamma^\gamma_{\beta\alpha}.$$

In the second expression $g = \det(g_{\alpha\beta})$ and $(g^{\alpha\beta})$ is the inverse matrix for $(g_{\alpha\beta})$.

We now apply this to the manifold V^3 and the function $f = \sqrt{-g_{00}}$ on this manifold. We expect the classical Poisson equation to take the form

$$\nabla^2 U = \nabla^2 (1 - \sqrt{-g_{00}}) = -4\pi\kappa\rho^*$$

or

$$\nabla^2 \sqrt{-g_{00}} = 4\pi\kappa\rho^* \tag{3-2}$$

which we shall call the *generalized Poisson equation.* We shall now look more closely at each side of this basic equation.

With respect to the left side we are confronted with the following purely mathematical problem: in a static M^4 compute the Laplacian of $\sqrt{-g_{00}}$ with respect to the spatial metric dl^2 of V^3 and relate this to the geometry of M^4. This was done by Levi-Civita in 1917 with the following result.

$$\boxed{\nabla^2 \sqrt{-g_{00}} = -R^0_0 \sqrt{-g_{00}}.} \tag{3-3}$$

In this equation $R_{ij} = R^k{}_{ikj}$ is the Ricci tensor (of M^4) and $R^l_i = g^{lj}R_{ij}$. A generalized version of this Levi-Civita formula is proven in Chapter 12 (Eqn. 12-23).

Comparing Levi-Civita's purely mathematical formula with the generalized Poisson equation leads us to conclude

$$R^0_0 \sqrt{-g_{00}} = -4\pi\kappa\rho^*, \tag{3-4}$$

giving us a relation between the density ρ^* and the geometry of M^4. We see already from this that ***we must expect space-time to be curved in the presence of matter!***

We should mention that while $\sqrt{-g_{00}}$ has a simple spatial Laplacian, given by Levi-Civita's formula, the Laplacian of g_{00} itself is more complicated, and comparison with the classical Poisson equation would be much more difficult. This is a most compelling reason for relating $\sqrt{-g_{00}}$ with the Newtonian potential rather than g_{00} or some other function of g_{00} (but see the discussion in Chapter 11).

The Density ρ^*

Now look at the right side $4\pi\kappa\rho^*$ of the generalized Poisson equation. Special relativity has taught us that *all* forms of energy possess mass; we expect then that ρ^* is the density of all mass-energy, not just mass.

Consider the fluid blob. We suppose that we are dealing with a "perfect fluid," so that the total mass-energy of the blob consists, on the Newtonian level, of two parts:

$$\text{total mass-energy} = (\text{rest energy of the particles}) + (\text{gravitational potential energy}),$$

where the first term includes not only the rest mass density of the particles but also the "compressional potential." Now let us express these quantities by their Newtonian values but modified by their appearance on the possible curved V^3 rather than Euclidean space. (The treatment that follows reverses one given by Tolman, 1934, pp. 248–250.)

Let ρ_0 be the rest energy density of the particles in the fluid blob $B \subset V^3$:

$$\text{rest energy} = \int_B \rho_0 \sqrt{g_V}\, dx$$

$$\text{potential energy} = -\int_B \tfrac{1}{2}\rho_0 U \sqrt{g_V}\, dx.$$

Here $\sqrt{g_V}\, dx = \sqrt{\det(g_{\alpha\beta})}\, dx^1\, dx^2\, dx^3$ is the volume form for V^3. For total mass-energy we have

$$\int_B \rho^* \sqrt{g_V}\, dx = \int_B \rho_0(1 - \tfrac{1}{2}U) \sqrt{g_V}\, dx.$$

We now rewrite (following Tolman) the right side for the case of a *small* spherical mass and consequently *weak* fields. U and the pressure p will be considered "small." Products of small terms will be omitted. $\sqrt{g_V}\, dx$ will be assumed to differ from the Euclidean volume element by a small quantity

$$\sqrt{g_V}\, dx \sim r^2 \sin\theta\, dr\, d\theta\, d\varphi.$$

From Equation 3-4 we expect that ρ^* contains $\sqrt{-g_{00}} = 1 - U$ as a factor. Then

$$\begin{aligned}\int_B \rho^* \sqrt{g_V}\, dx &= \int_B \frac{\rho_0(1 - \tfrac{1}{2}U)}{1 - U} \sqrt{-g_{00}} \sqrt{g_V}\, dx \\ &\sim \int_B \rho_0(1 + \tfrac{1}{2}U) \sqrt{-g_{00}} \sqrt{g_V}\, dx.\end{aligned} \tag{3-5}$$

On the other hand, we have for a spherical ball of radius r_0

$$\begin{aligned}\tfrac{1}{2} \int_B \rho_0 U \sqrt{-g_{00}} \sqrt{g_V}\, dx &\sim \tfrac{1}{2} \int_B \rho_0 U r^2 \sin\theta\, dr\, d\theta\, d\varphi \\ &= -\,(\text{Newtonian potential energy of } B) \\ &= \int_B \frac{\kappa M_r dM_r}{r} = \int_B r\, \frac{\kappa M_r dM_r}{r^2},\end{aligned}$$

where M_r is the total mass contained in the ball of radius r. But, roughly speaking, the gravitational force $\kappa r^{-2} M_r \, dM_r$ on the shell dM_r is balanced by the hydrostatic force due to pressure

$$\begin{aligned}\int_B r \frac{\kappa M_r \, dM_r}{r^2} &= -\int_B r \cdot 4\pi r^2 \, dp \\ &= -4\pi r^3 p \Big]_0^{r_0} + \int_0^{r_0} 3p \cdot 4\pi r^2 \, dr \\ &\sim \int_B 3p \sqrt{-g_{00}} \sqrt{g_V} \, dx,\end{aligned}$$

since $p = 0$ on the surface $r = r_0$ of the fluid ball. Thus from Equation 3-5,

$$\int_B \rho^* \sqrt{g_V} \, dx \sim \int_B (\rho_0 + 3p) \sqrt{-g_{00}} \sqrt{g_V} \, dx, \tag{3-6}$$

and we are led, at least in this case of a weak field, to propose the following heuristic definition:

$$\rho^* \equiv (\rho_0 + 3p) \sqrt{-g_{00}}. \tag{3-7}$$

Note that Equation 3-7 does not *follow* from Equation 3-6, since B is not arbitrary. We must integrate over the entire fluid ball and take advantage of the vanishing of p on the boundary.

Einstein's Equations

As defined in Equation 3-7, ρ^* replaces the classical notion of mass density! Comparing with Equation 3-4 then yields

$$-R_0^0 = 4\pi\kappa(\rho_0 + 3p) \tag{3-8}$$

and the generalized Poisson equation (Eqn. 3-2) becomes

$$\nabla^2 \sqrt{-g_{00}} = 4\pi\kappa(\rho_0 + 3p) \sqrt{-g_{00}}. \tag{3-9}$$

Equation 3-8 describes how one component of the Ricci tensor of M^4 is determined by the characteristics ρ_0 and p of the fluid, at least in this static case. To understand the general case we shall also write the right side of Equation 3-8 as the $\binom{0}{0}$ component of a tensor. The result will then suggest a tensor equation for M^4. The tensor in question is the *stress-energy-momentum tensor.*

This tensor was introduced into special relativity during the period 1908–1911. For the electromagnetic field it generalizes the electromagnetic stress tensor of Faraday and Maxwell, and is discussed in Chapter 10 (p. 118ff). For the case of a fluid it generalizes the classic Cauchy stress tensor (see Chapter 6, p. 64ff). For the present the reader may simply assume that there is a physical tensor that, in the case of a perfect fluid, has components

$$T^{ij} = (\rho_0 + p)u^i u^j + pg^{ij}.$$

Here u is the unit velocity 4-vector of the fluid flow

$$u^i = \frac{dx^i}{d\tau}\,.$$

In our static case the fluid is spatially at rest, i.e., $dx^\alpha/d\tau = 0$ for $\alpha = 1, 2, 3$, and so the only nonvanishing component of u is

$$u^0 = \frac{dt}{d\tau} = \frac{1}{\sqrt{-g_{00}}}\,.$$

Then, for our metric of type $ds^2 = g_{00}\,dt^2 + dl^2$,

$$\begin{aligned} T^0_0 = g_{0j}T^{0j} &= (\rho_0 + p)g_{0j}u^0u^j + pg_{0j}g^{0j} \\ &= (\rho_0 + p)g_{00}u^0u^0 + pg_{00}g^{00} \\ &= -(\rho_0 + p) + p = -\rho_0. \end{aligned}$$

Also $T^j_i = (\rho_0 + p)u_iu^j + p\delta^j_i$, and so for trace we have

$$T \equiv T^i_i = (\rho_0 + p)\langle u, u\rangle + 4p = -(\rho_0 - 3p),$$

since $\langle u, u\rangle = g_{ik}u^iu^k = -1$. Consequently,

$$\rho_0 + 3p = T - 2T^0_0 = T^1_1 + T^2_2 + T^3_3 - T^0_0 \tag{3-10}$$

and Equation 3-8 becomes

$$R^0_0 = 8\pi\kappa(T^0_0 - \tfrac{1}{2}T). \tag{3-11}$$

R^0_0 is the $\binom{0}{0}$ component of the Ricci tensor of M^4. $T^0_0 - \frac{1}{2}T$ is the $\binom{0}{0}$ component of the physical tensor whose $\binom{j}{i}$ component is $T^j_i - \frac{1}{2}\delta^j_i T$. We are led to propose that the following tensor equation holds:

$$\boxed{R^j_i = 8\pi\kappa(T^j_i - \tfrac{1}{2}\delta^j_i T)} \tag{3-12}$$

Equations 3-12 were announced by Einstein in November 1915. They were also obtained independently by Hilbert using an entirely different approach from that of Einstein or the above. (Hilbert's paper actually appeared several days before Einstein's, but although independent of his, it was prompted by Einstein's earlier incomplete attempts. For the history see Mehra, 1974, and Lanczos, 1974. For Hilbert's "variational method" see Adler, Bazin, and Schiffer, 1975.)

Taking the trace of both sides of Equation 3-12 we get for scalar curvature of M^4

$$R = R^i_i = 8\pi\kappa(T^i_i - \tfrac{1}{2}\delta^i_i T) = -8\pi\kappa T, \tag{3-13}$$

and we have an equivalent form of the Einstein equations

$$R^j_i - \tfrac{1}{2}\delta^j_i R = 8\pi\kappa T^j_i$$

or, in purely covariant form

$$R_{ij} - \tfrac{1}{2}g_{ij}R = 8\pi\kappa T_{ij}. \tag{3-14}$$

We have been led to these equations by a very simple, static situation. *We now postulate that these equations are true in general, for matter and energy undergoing the most diverse motions.* Our heuristic derivation is strengthened by the following remark. Earlier results of Weyl and Cartan culminated in Lovelock's statement (1972) that if we seek a tensor equation of the form $A^{ij} = T^{ij}$, where the components (A^{ij}) involve the metric tensor (g_{kl}) and its first and second derivatives (thus assuring second-order partial differential equations generalizing the Poisson equation), and if (A^{ij}) is to have vanishing divergence $A^{ij}_{;j} = 0$, then the equation must be of the form

$$a(R^{ij} - \tfrac{1}{2}g^{ij}R) + bg^{ij} = 8\pi\kappa T^{ij},$$

where a and b are constants. The need for $A^{ij}_{;j} = 0$ will be better understood after our discussion of the stress-energy-momentum tensor in Chapter 7. Einstein's choice is then $a = 1, b = 0$.

To understand the consequences of the Einstein equations we shall first express them in a more geometric form in the next chapter.

4

The Geometry of Einstein's Equations

Curvature in a Pseudo-Riemannian M^4

We shall define here the geometric notions associated with a pseudo-Riemannian manifold (and in particular the space-time manifold M^4). There are troublesome sign differences from the case of a Riemannian manifold. Even in the case of a Riemannian manifold there are several sign conventions in the literature. Our sign conventions are the same as in the book of Misner, Thorne, and Wheeler (1973).

Let ∇ be the covariant differential operator; if X is a vector at $p \in M$ and Y is a vector field defined near p, then $\nabla_X Y$ is the covariant derivative of Y with respect to X. In terms of any basis $\{e_i\}$ of the tangent vectors near p,

$$\nabla_X Y = Y^i{}_{;k} X^k e_i.$$

In terms of a coordinate basis $e_i = \partial/\partial x^i$,

$$Y^i{}_{;k} = \frac{\partial Y^i}{\partial x^k} + \Gamma^i{}_{kj} Y^j$$

where $\Gamma^i{}_{kj}$ are the Christoffel symbols

$$\Gamma^i{}_{kj} = \frac{1}{2} g^{il} \left(\frac{\partial g_{lk}}{\partial x^j} + \frac{\partial g_{lj}}{\partial x^k} - \frac{\partial g_{kj}}{\partial x^l} \right).$$

The Lie (or commutator) bracket of a pair of vector fields X, Y is the vector field

$$\begin{aligned} [X, Y] &= \nabla_X Y - \nabla_Y X \\ &= (Y^i{}_{;k} X^k - X^i{}_{;k} Y^k) e_i. \end{aligned}$$

If the e_i are coordinate vectors, then the covariant derivatives can be replaced by ordinary derivatives (see Chapter 6, p. 57ff), i.e., we have the usual expression

$$[X, Y] = \left(\frac{\partial Y^i}{\partial x^k} X^k - \frac{\partial X^i}{\partial x^k} Y^k \right) \frac{\partial}{\partial x^i}.$$

The curvature transformation $\mathsf{R}(X, Y)$ for a pair of vectors at $p \in M$ is the linear transformation of M_p into itself, defined by

$$\mathsf{R}(X_p, Y_p)Z_p = \nabla_{X_p} \nabla_Y Z - \nabla_{Y_p} \nabla_X Z - \nabla_{[X, Y]_p} Z \tag{4-1}$$

where X, Y, Z are any extensions of X_p, Y_p, Z_p to vector fields. In terms of a basis, components $(R^l{}_{kij})$ are defined by

$$\mathsf{R}(e_i, e_j) e_k = R^l{}_{kij} e_l.$$

In a coordinate basis

$$R^l{}_{kij} = \frac{\partial \Gamma^l{}_{kj}}{\partial x^i} - \frac{\partial \Gamma^l{}_{ki}}{\partial x^j} + \Gamma^h{}_{kj} \Gamma^l{}_{hi} - \Gamma^h{}_{ki} \Gamma^l{}_{hj}.$$

Curvature measures noncommutativity of covariant derivatives. In a coordinate basis, Equation 4-1 yields the *Ricci identities*

$$Z^l{}_{;ji} - Z^l{}_{;ij} = Z^k R^l{}_{kij}.$$

A vector ξ is a *unit vector* if

$$\varepsilon_\xi \equiv \langle \xi, \xi \rangle = \|\xi\|^2 = g_{ik}\xi^i\xi^k$$

is ± 1. ε_ξ is the *indicator* of the unit vector ξ.

Let $\{e_0, e_1, e_2, e_3\}$ be an *orthonormal* basis of space-time M^4 at an event p, that is,

$$\langle e_i, e_j \rangle = \begin{cases} -1 & \text{if } i = j = 0 \\ +1 & \text{if } i = j \neq 0 \\ \ \ 0 & \text{if } i \neq j. \end{cases}$$

For indicators we shall put

$$\varepsilon_i = \varepsilon_{e_i}$$

and so $\varepsilon_0 = -1$, $\varepsilon_\alpha = +1$, $\alpha = 1, 2, 3$.

If we deal with a general (not necessarily orthonormal) frame, $\langle e_i, e_j \rangle = g_{ij}$ and we can always write $X = X^i e_i$, where $X^i = g^{ij}\langle X, e_j \rangle$. When an orthonormal frame is used we have $X = \sum_i \varepsilon_i \langle X, e_i \rangle e_i$. Also, in a general frame we have

$$\langle \mathsf{R}(e_i, e_j)e_k, e_m \rangle = R^l{}_{kij}\langle e_l, e_m \rangle = R^l{}_{kij}g_{lm} = R_{mkij}.$$

Many other placements of the indexes in the Riemann tensor will be found in the literature!

The volume form ω^4 for the orientation defined by the orthonormal frame e_0, e_1, e_2, e_3 is defined by the property $\omega^4(e_0, e_1, e_2, e_3) = 1$. If x^0, x^1, x^2, x^3 are local coordinates defining the same orientation, then $\dfrac{\partial}{\partial x^i} = \Sigma_j \varepsilon_j \left\langle \dfrac{\partial}{\partial x^i}, e_j \right\rangle e_j$ and

$$\omega^4 = \omega^4\left(\frac{\partial}{\partial x^0}, \frac{\partial}{\partial x^1}, \frac{\partial}{\partial x^2}, \frac{\partial}{\partial x^3}\right) dx^0 \wedge dx^1 \wedge dx^2 \wedge dx^3,$$

where $\omega^4\left(\dfrac{\partial}{\partial x^0}, \dfrac{\partial}{\partial x^1}, \dfrac{\partial}{\partial x^2}, \dfrac{\partial}{\partial x^3}\right) = \det\left\langle \dfrac{\partial}{\partial x^i}, \varepsilon_j e_j \right\rangle \cdot$ We then have

$$g_{ik} = \left\langle \frac{\partial}{\partial x^i}, \frac{\partial}{\partial x^k} \right\rangle = \sum_j \left\langle \frac{\partial}{\partial x^i}, \varepsilon_j e_j \right\rangle \left\langle e_j, \frac{\partial}{\partial x^k} \right\rangle.$$

Thus

$$g = \det(g_{ik}) = \det\left\langle \frac{\partial}{\partial x^i}, \varepsilon_j e_j \right\rangle \det\left\langle e_j, \frac{\partial}{\partial x^k} \right\rangle$$
$$= -\left[\det\left\langle \frac{\partial}{\partial x^i}, \varepsilon_j e_j \right\rangle\right]^2,$$

since $\det\left\langle \varepsilon_j e_j, \frac{\partial}{\partial x^k} \right\rangle = -\det\left\langle e_j, \frac{\partial}{\partial x^k} \right\rangle \cdot$ Finally then

$$\omega^4 = \sqrt{-g}\, dx^0 \wedge dx^1 \wedge dx^2 \wedge dx^3.$$

Let X and Y be vectors at $p \in M$. The "square" of the area of the parallelogram spanned by X and Y is defined by

$$\|X \wedge Y\|^2 = \|X\|^2 \|Y\|^2 - \langle X, Y \rangle^2,$$

which need not be positive. In particular, for the orthonormal frame $\{e_i\}$

$$\|e_i \wedge e_j\|^2 = \varepsilon_i \varepsilon_j \quad \text{when } i \neq j.$$

The *Riemann sectional curvature* $\mathsf{K}(X \wedge Y)$ for the 2-plane spanned by X and Y is defined, when $\|X \wedge Y\|^2 \neq 0$, by

$$\mathsf{K}(X \wedge Y) = \frac{\langle \mathsf{R}(X, Y)Y, X \rangle}{\|X \wedge Y\|^2} = \mathsf{K}(Y \wedge X). \tag{4-2}$$

(This depends only on the 2-plane, not the spanning set X, Y.) If X and Y are dependent we shall use the convention $\mathsf{K}(X \wedge Y) = 0$. In terms of an orthonormal basis,

$$\mathsf{K}(e_i \wedge e_j) = \varepsilon_i \varepsilon_j \langle \mathsf{R}(e_i, e_j)e_j, e_i \rangle = \varepsilon_i \varepsilon_j R_{ijij}. \tag{4-3}$$

(There is no sum implied here since all indexes are subscripts.) The geometric interpretation is

$$\mathsf{K}(e_i \wedge e_j) = \text{the Gaussian curvature at } p \text{ of the two-dimensional manifold formed by all the geodesics through } p \text{ that are tangent to the 2-plane } e_i \wedge e_j. \tag{4-4}$$

The Ricci tensor $(R_{ij}) = (R^k{}_{ikj})$ defines a quadratic form; for a vector ξ

$$\mathsf{Ric}(\xi, \xi) = R_{ij}\xi^i \xi^j.$$

This quadratic form has a definition independent of components:

$$\mathsf{Ric}(\xi, \xi) = \text{the trace of the linear transformation}$$

$$\mathsf{A}_\xi : \eta \to \mathsf{R}(\eta, \xi)\xi.$$

We can then see the geometric significance of the Ricci quadratic form. $\mathsf{A}_\xi(e_i) = \mathsf{R}(e_i, \xi)\xi = \sum_j \varepsilon_j \langle \mathsf{R}(e_i, \xi)\xi, e_j \rangle e_j$, and so

$$\mathsf{Ric}(\xi, \xi) = \sum_i \varepsilon_i \langle \mathsf{R}(e_i, \xi)\xi, e_i \rangle.$$

Thus, if ξ is one of the basis vectors e_0, e_1, e_2, or e_3,

$$\mathsf{Ric}(e_j, e_j) = \varepsilon_j \sum_i \mathsf{K}(e_j \wedge e_i) = \varepsilon_j \sum_{i \neq j} \mathsf{K}(e_j \wedge e_i), \tag{4-5}$$

and so $\mathsf{Ric}(e_j, e_j)$ is essentially a sum of sectional curvatures.

To interpret the scalar curvature let ric be the linear transformation (whose matrix is (R^j_i)) associated with the form Ric, i.e., $\langle \mathsf{ric}(e_i), e_i \rangle = \mathsf{Ric}(e_i, e_i)$. Then $R = R^i_i = \text{trace ric}$. Now $\mathsf{ric}(e_i) = \Sigma_j \varepsilon_j \langle \mathsf{ric}(e_i), e_j \rangle e_j$, and so

$$\begin{aligned} R &= \text{trace ric} = \sum_i \varepsilon_i \langle \mathsf{ric}(e_i), e_i \rangle = \sum_i \varepsilon_i \, \mathsf{Ric}(e_i, e_i) \\ &= \sum_i \varepsilon_i^2 \sum_{k \neq i} \mathsf{K}(e_i \wedge e_k) = \sum_{\substack{i,k \\ i \neq k}} \mathsf{K}(e_i \wedge e_k) \end{aligned} \tag{4-6}$$

is also a sum of sectional curvatures.

The Einstein Tensor G_{ij}

Define the Einstein tensor

$$G_{ij} = R_{ij} - \tfrac{1}{2} g_{ij} R \tag{4-7}$$

with its associated quadratic form

$$\mathsf{G}(\xi, \xi) = G_{ij}\xi^i\xi^j = \mathsf{Ric}(\xi, \xi) - \tfrac{1}{2}\langle \xi, \xi \rangle R.$$

In particular, for our orthonormal basis

$$\begin{aligned}\mathsf{G}(e_i, e_i) &= \mathsf{Ric}(e_i, e_i) - \tfrac{1}{2}\varepsilon_i R \\ &= \varepsilon_i \sum_{j\neq i} \mathsf{K}(e_i \wedge e_j) - \tfrac{1}{2}\varepsilon_i \sum_{k,j} \mathsf{K}(e_k \wedge e_j) \\ &= \varepsilon_i \left[\sum_{j\neq i} \mathsf{K}(e_i \wedge e_j) - \sum_{k<j} \mathsf{K}(e_k \wedge e_j)\right] \\ &= -\varepsilon_i \sum_{\substack{k\neq i\neq j \\ k<j}} \mathsf{K}(e_k \wedge e_j).\end{aligned}$$

If we let $e_i^\perp$ be any plane of the form $e_j \wedge e_k$ for $j \neq i$, $k \neq i$, we have

$$\mathsf{G}(e_i, e_i) = -\varepsilon_i \Sigma \mathsf{K}(e_i^\perp).$$

Thus $-\varepsilon_i \mathsf{G}(e_i, e_i)$ is the sum of the sectional curvatures of the three coordinate 2-planes not containing e_i. For example,

$$\mathsf{G}(e_0, e_0) = \mathsf{K}(e_1 \wedge e_2) + \mathsf{K}(e_1 \wedge e_3) + \mathsf{K}(e_2 \wedge e_3).$$

We now return to the Einstein equations $G_{ij} = 8\pi\kappa T_{ij}$, which can be written $\mathsf{G}(\xi, \xi) = 8\pi\kappa \mathsf{T}(\xi, \xi)$ for any vector ξ, where $\mathsf{T}(\xi, \xi) = T_{ij}\xi^i\xi^j$. We then have the following version of the Einstein equations relating stress-energy-momentum to Riemann sectional curvatures

$$\boxed{\Sigma \mathsf{K}(e_i^\perp) = -8\pi\kappa\varepsilon_i \mathsf{T}(e_i, e_i)} \tag{4-8}$$

Consider, for example, the case of a perfect fluid, whose stress-energy-momentum tensor is (as we shall see in Chapter 7) in its various forms

$$\begin{aligned}T^{ij} &= (\rho_0 + p)u^i u^j + p g^{ij} \\ T_i^j &= (\rho_0 + p)u_i u^j + p\delta_i^j \\ T_{ij} &= (\rho_0 + p)u_i u_j + p g_{ij},\end{aligned}$$

where u is the unit velocity 4-vector of the fluid and ρ_0 is the "rest density." Then, since $\langle u, u\rangle = -1$,

$$\begin{aligned}T_i^j u^i &= (\rho_0 + p)u_i u^j u^i + p\delta_i^j u^i \\ &= -(\rho_0 + p)u^j + p u^j = -\rho_0 u^j\end{aligned}$$

i.e., the unit velocity vector of the flow is a time-like eigenvector of the stress-energy-momentum tensor with eigenvalue $-\rho_0$. On the other hand, if v is any unit vector orthogonal to u, then $\langle u, v\rangle = 0$,

$$T_i^j v^i = p\delta_i^j v^i = pv^j,$$

and so v is automatically a space-like eigenvector with eigenvalue p, the pressure. Then $\mathsf{T}(u, u) = (\rho_0 + p) - p = \rho_0$ and $\mathsf{T}(v, v) = pg_{ij}v^iv^j = p$, and Einstein's equations geometrically state, when we choose $e_0 = u$,

$$\left\{\begin{aligned} \Sigma\mathsf{K}(e_0^\perp) &= 8\pi\kappa\mathsf{T}(e_0, e_0) = 8\pi\kappa\rho_0 \\ \Sigma\mathsf{K}(e_\alpha^\perp) &= -8\pi\kappa\mathsf{T}(e_\alpha, e_\alpha) = -8\pi\kappa p \end{aligned}\right\} \tag{4-9}$$

for $\alpha = 1, 2, 3$.

The Gauss Equations in M^4

We have interpreted Einstein's equations as equating the quadratic form of stress-energy-momentum essentially with a sum of sectional curvatures of M^4. For our purposes it will be more useful to bring in directly the geometry of the spatial sections V_t^3 of M^4 via the so-called Gauss equations of a hypersurface of M^4. For this purpose first recall the simpler situation of a two-dimensional surface sitting in Euclidean space R^3.

Let $V^2 \subset R^3$ be a surface in R^3 with a continuous field of *unit* normal vectors N. V^2 can be considered a Riemannian manifold with metric induced from the Euclidean metric of R^3. As such it has only one Riemannian sectional curvature; in terms of the Riemann tensor it is $K_V = R^{12}{}_{12}$. On the other hand, V^2 as a submanifold of R^3 has a *second fundamental form* b, defined as follows. Let V_p^2 be the two-dimensional tangent plane to V at p; b is the linear transformation b: $V_p^2 \to V_p^2$, defined by

$$\mathsf{b}(X) = -\nabla_X N.$$

Here ∇_X is the covariant (in this case ordinary) derivative with respect to the tangent vector X (since N has constant length, $\nabla_X N$ is automatically tangent to V). The linear transformation b is self-adjoint and thus has two real eigenvalues κ_1, κ_2, the "principal normal curvatures." (The corresponding eigendirections can be chosen to be orthogonal and are the "principal directions" at p.) The *Gauss curvature* of V at p is the product of the principal curvatures, and Gauss's *theorema egregium* states

$$K_V = \kappa_1\kappa_2 = \det(\mathsf{b}).$$

All this can be vastly generalized. For the present we shall be interested only in the following case.

Let $V^3 \subset M^4$ be a three-dimensional submanifold of space-time M^4, and let N be a continuous field of *unit* vectors to V^3. The second fundamental form $\mathsf{b} : V^3_p \to V^3_p$ is again defined by $\mathsf{b}(X) = -\nabla_X N$, where covariant differentiation takes place in M^4. b is again a self-adjoint linear transformation but since the scalar product in V^3_p need not be positive definite, the eigenvalues of b need not be real. (The usual procedure of maximizing $\langle \mathsf{b}(X), X \rangle$ subject to $\langle X, X \rangle = 1$ might fail if the "sphere" $\{\langle X, X \rangle = 1\}$ is not compact.) Assume, however, that b does have three real eigenvalues; in particular this will happen if V^3 is space-like (i.e., if tangent vectors are space-like), for the induced scalar product is positive definite. The eigendirections can again be chosen to be orthogonal. Let X and Y be unit, orthogonal eigenvectors corresponding to eigenvalues κ_X and κ_Y. Then the generalization of Gauss's *theorema egregium* is the *Gauss equation*

$$\mathsf{K}(X \wedge Y) = \mathsf{K}_V(X \wedge Y) - \varepsilon_N \kappa_X \kappa_Y. \tag{4-10}$$

Here $\mathsf{K}(X \wedge Y)$ is the Riemann sectional curvature of M^4, $\mathsf{K}_V(X \wedge Y)$ is the Riemann sectional curvature of V^3 using the *induced metric* on V^3, and $\varepsilon_N = \langle N, N \rangle$ is the indicator of the unit normal. The proof of this Gauss equation proceeds as in the Riemannian case: from Equations 4-1 and 4-2 one computes

$$\mathsf{K}(X \wedge Y) = \mathsf{K}_V(X \wedge Y) - \varepsilon_N \varepsilon_X \varepsilon_Y \{\mathsf{B}(X, X)\mathsf{B}(Y, Y) - \mathsf{B}^2(X, Y)\}, \tag{4-11}$$

where $\mathsf{B}(W, Z) = \langle \mathsf{b}(W), Z \rangle = \mathsf{B}(Z, W)$ is the bilinear form associated with the linear transformation b.

A Geometric Form of Einstein's Equations

Let V^3 be a space-like hypersurface, let $e_0 = N$ be a unit normal field, and let e_1, e_2, e_3 be unit orthogonal tangent vectors to V^3. Then from Equation 4-8,

$$8\pi\kappa \mathsf{T}(e_0, e_0) = \mathsf{K}(e_1 \wedge e_2) + \mathsf{K}(e_1 \wedge e_3) + \mathsf{K}(e_2 \wedge e_3).$$

Choose e_1, e_2, e_3 to be principal directions on V^3 corresponding to principal curvatures $\kappa_1, \kappa_2, \kappa_3$. Then

$$\begin{aligned} 8\pi\kappa \mathsf{T}(e_0, e_0) &= \mathsf{K}_V(e_1 \wedge e_2) + \mathsf{K}_V(e_1 \wedge e_3) + \mathsf{K}_V(e_2 \wedge e_3) \\ &\quad + \kappa_1\kappa_2 + \kappa_1\kappa_3 + \kappa_2\kappa_3 \\ &= \tfrac{1}{2} R_{V^3} + \kappa_1\kappa_2 + \kappa_1\kappa_3 + \kappa_2\kappa_3, \end{aligned} \tag{4-12}$$

where R_{V^3} is the scalar curvature of V^3 in its own (induced) Riemannian metric.

$H = \mathrm{tr}(\mathsf{b}) = \kappa_1 + \kappa_2 + \kappa_3$ is called the *mean* curvature of $V^3 \subset M^4$. We shall write

$$\mathrm{tr}(\mathsf{b} \wedge \mathsf{b}) \equiv \kappa_1\kappa_2 + \kappa_1\kappa_3 + \kappa_2\kappa_3$$

(because this *is* the trace of the natural extension of b to a linear transformation of bi-vectors). Einstein's equations then yield

$$8\pi\kappa \mathsf{T}(e_0, e_0) = \tfrac{1}{2}R_{V^3} + \mathrm{tr}(\mathsf{b} \wedge \mathsf{b}),$$

a form of Einstein's equations that has been emphasized by Wheeler. Note that if we had started with a general V^3 with unit normal ξ and orthonormal eigenvectors for b, we would have

$$\boxed{8\pi\kappa \mathsf{T}(\xi, \xi) = \mathsf{G}(\xi, \xi) = -\varepsilon_\xi \cdot \tfrac{1}{2}R_{V^3} + \mathrm{tr}(\mathsf{b} \wedge \mathsf{b}),} \tag{4-13}$$

which is our final geometric form of Einstein's equations. The second of these equalities, which is entirely geometric, has the following interpretation (a consequence of the Gauss equation). Let ξ be any unit vector at $p \in M^4$. Take any V^3 that is perpendicular to ξ at p. This V^3 has a scalar curvature R_V at p and a type of "curvature" $\mathrm{tr}(\mathsf{b} \wedge \mathsf{b})$ at p. While both of these "curvatures" depend on how V^3 "curves" in M^4, the expression $-\varepsilon_\xi \cdot \tfrac{1}{2}R_V + \mathrm{tr}(\mathsf{b} \wedge \mathsf{b})$ is *independent* of this situation; it depends only on the normal ξ at p and equals the Einstein quadratic expression $\mathsf{G}(\xi, \xi)$. In Minkowski space this expression vanishes. It will not vanish, in general relativity, when energy or stresses are present.

Some final remarks. First, the right-hand equality of Equation 4-13 holds in the situation $V^n \subset M^{n+1}$, for $n \geq 2$, that is,

$$\mathsf{G}(\xi, \xi) = -\varepsilon_\xi \tfrac{1}{2}R_{V^n} + \mathrm{tr}(\mathsf{b} \wedge \mathsf{b}). \tag{4-14}$$

Second, when V^2 is any Riemannian surface, it is common to call $K_V = \tfrac{1}{2}R_V = \mathsf{R}^{12}{}_{12}$ the Gauss curvature of V^2. This is the usage of the term that appears in the geometric interpretation of $\mathsf{K}(e_i \wedge e_j)$ given by Equation 4-4.

Third, we have seen in Equation 4-9 that when ξ is the velocity vector for a perfect fluid, $\mathsf{T}(\xi, \xi)$ reduces to the ordinary (rest) density of the fluid, and from Equations 4-9 and 4-13 that $\mathsf{T}(\xi, \xi)$ is directly related to certain curvatures associated with the 3-plane orthogonal to ξ. For these reasons we are led to make the following definition. If ξ is any time-like unit vector,

$$\rho \equiv \mathsf{T}(\xi, \xi) \tag{4-15}$$

will be called the (*curvature*) *energy density*. Consider, for example, fluid motion. Let an observer set up a "Gaussian" coordinate system, in which the metric takes the form

$$ds^2 = -dt^2 + g_{\alpha\beta}\, dx^\alpha\, dx^\beta.$$

This can always be accomplished by taking a fixed spatial hypersurface and taking for time-lines the geodesics, leaving this hypersurface orthogonally, and using proper time for time parameter ("Gauss's lemma"). The unit velocity vector for the world lines of the fluid is given by

$$u = \frac{dt}{d\tau}\frac{\partial}{\partial t} + \frac{dx^\alpha}{d\tau}\frac{\partial}{\partial x^\alpha}$$

where τ is proper time *along the fluid world lines*. But $d\tau^2 = dt^2 - g_{\alpha\beta}\, dx^\alpha\, dx^\beta$ yields $dt/d\tau = \gamma$, where

$$\gamma = (1 - |\mathbf{v}|^2)^{-1/2}$$

and where $\mathbf{v} = dx^\alpha/dt\ \partial/\partial x^\alpha$ is the classical velocity vector. Consequently, letting $\xi = \partial/\partial t$,

$$\begin{aligned} \mathsf{T}(\xi, \xi) &= (\rho_0 + p)\langle u, \xi\rangle^2 - p \\ &= (\rho_0 + p)\gamma^2 - p \\ &= \rho_0\gamma^2 + p(\gamma^2 - 1). \end{aligned}$$

The dominant term $\rho_0\gamma^2$ can be understood in terms of special relativity. A fluid particle whose rest mass is m_0 will be seen by the observer to have mass $m = m_0\gamma$ (see Eqn. 1-10). The volume of the particle will appear to be altered by the factor γ^{-1} due to the Lorentz-Fitzgerald contraction (Eqn. 1-3) in the single direction of motion. Thus if ρ_0 is the rest mass density, *the density as measured by the observer will be* $\rho_0\gamma^2$. (It is this factor of γ^2 from the Lorentz transformation (Eqn. 1-2) that suggests that density of mass is a component of a *second*-rank tensor.)

The situation in which ξ is the unit normal field to a *totally geodesic* space-like hypersurface is especially striking. Recall that V^3 is totally geodesic if every geodesic of V^3 (in the induced metric) is also a geodesic of M^4. It is equivalent to say that the second fundamental form b of V^3 vanishes identically. From Equation 4-13 we then have for the scalar curvature of V^3

$$\boxed{\begin{gathered} R_{V^3} = 16\pi\kappa\rho \\ V^3 \quad \text{totally geodesic,} \end{gathered}} \tag{4-16}$$

a particularly beautiful expression of Einstein's equation.

5

The Schwarzschild Solution

Schwarzschild Coordinates

The Einstein equations give a relation between matter (T_{ij}) and the Ricci tensor (R_{ij}) of space-time, not the full Riemann tensor. For example, a region of space-time is said to be *empty* if $T_{ij} = 0$ there. The Einstein equations, $R_{ij} = 8\pi\kappa(T_{ij} - \frac{1}{2}g_{ij}T)$, then say that the Ricci tensor vanishes in this region; the region need not be flat but can be curved because of matter elsewhere.

The simplest and most important case is concerned with a *spherically symmetric static mass distribution,* like an idealized sun in an otherwise empty universe. As in Chapter 2, we would expect the static, local line element $ds^2 = g_{00}(x)dt^2 + dl^2$, where both the spatial metric dl^2 and g_{00} are independent of time $t = x^0$. We may try to find such a universe of the form $M^4 = R \times R^3$, with matter centered at the origin of R^3 and it is natural to introduce spherical coordinates in R^3, (r, θ, φ). We may *not* assume that the spatial metric is the flat Euclidean one $dr^2 + r^2(d\theta^2 +$

$\sin^2\theta\, d\varphi^2$) but spherical symmetry does imply that the 2-spheres, $r =$ constant, in the spatial sections V_t^3 carry a metric of constant Gauss curvature, the constant depending on r. We shall *normalize* the coordinate r by demanding that the 2-spheres S_r^2, $r =$ constant, have Gauss curvature $1/r^2$, just like the Euclidean sphere of radius r. Thus S_r^2 has area $4\pi r^2$. It is important to realize that this is merely a convenient normalization of the r-coordinate in this spherically symmetric situation. The metric of space-time is then of the form

$$\begin{cases} ds^2 = g_{00}(r)\, dt^2 + g_{rr}(r)\, dr^2 + r^2\, d\Omega^2 \\ d\Omega^2 = d\theta^2 + \sin^2\theta\, d\varphi^2, \end{cases} \tag{5-1}$$

$d\Omega^2$ being the standard metric of the unit 2-sphere in Euclidean 3-space. We now proceed to the determination of g_{00} and g_{rr}.

Embedding the Spatial Section

Each spatial section V_t^3 of the space-time manifold M^4 is an isometric copy of $V^3 = V_0^3$, since the metric (Eqn. 5-1) is static, and carries the metric $dl^2 = g_{rr}\, dr^2 + r^2\, d\Omega^2$. We shall try to realize the metric of V^3 (not M^4) by embedding V^3 as a three-dimensional *manifold of revolution* $\tilde{V}^3$ in R^4, as in Figure 5-1. In this figure $\tilde{V}^3$ is to be defined analytically by a single equation $w = w(r, \theta, \varphi) = w(r)$ in the single "cylindrical" radius r.

The original spatial section $V = V_0^3$ of M^4 is clearly the fixed set of the *isometry* $(t, \mathbf{x}) \to (-t, \mathbf{x})$ of M^4 into itself. But the fixed set of any isometry is automatically totally geodesic (since isometries take geodesics into geodesics and a geodesic is determined by its tangent at a single point). Thus by Equation 4-16,

$$R_{V^3} = 16\pi\kappa\rho = 16\pi\kappa \mathsf{T}(\xi, \xi),$$

where ξ is the unit normal to V_0^3. Since we wish the embedded version $\tilde{V}^3$ of V^3 to be isometric with V^3, we are to choose $w = w(r)$ so that

$$R_{\tilde{V}^3} = 16\pi\kappa\rho(r). \tag{5-2}$$

Now the metric on V^3 is $dl^2 = g_{rr}(r)\, dr^2 + r^2\, d\Omega^2$, while that on $\tilde{V}^3$ is induced from the Euclidean metric $dw^2 + dr^2 + r^2\, d\Omega^2$ of R^4, hence

$$g_{rr}(r) = 1 + \left(\frac{dw}{dr}\right)^2. \tag{5-3}$$

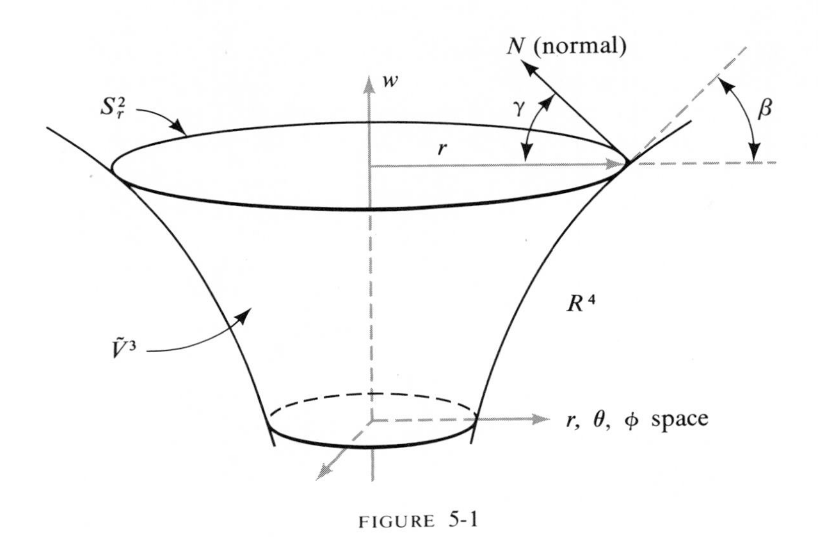

FIGURE 5-1

The embedded $\tilde{V}^3$ is a submanifold of the "Riemannian" R^4, which of course is flat; in particular, the Einstein tensor G_{R^4} of R^4 vanishes. If we let N be the unit normal to $\tilde{V}^3 \subset R^4$, and apply the Gauss equation (Eqn. 4-13) to $\tilde{V} \subset R^4$, we get

$$0 = \mathsf{G}_{R^4}(N, N) = -\varepsilon_N \cdot \tfrac{1}{2} R_{\tilde{V}^3} + k_1 k_2 + k_1 k_3 + k_2 k_3,$$

where k_1, k_2, k_3 are the principal curvatures of $\tilde{V}^3 \subset R^4$, and so we have

$$R_{V^3} = R_{\tilde{V}^3} = 2(k_1 k_2 + k_1 k_3 + k_2 k_3). \tag{5-4}$$

Let us now determine these principal normal curvatures. One of the principal directions will be a "line of longitude" and the other two will be tangent to the 2-spheres S_r^2 defined by r = constant (i.e., w = constant).

Now a great circle on S_r^2 has a radius of curvature r, curvature $1/r$, and thus normal curvature $1/r \cos \gamma = 1/r \sin \beta$. But $\sin \beta = (dw/dr) \cdot [1 + dw/dr)^2]^{-1/2}$ and so $\sin \beta = dw/dr\ g_{rr}{}^{-1/2}(r) = [1 - 1/g_{rr}(r)]^{1/2}$. We have found two principal curvatures:

$$k_1 = k_2 = \frac{1}{r}[1 - 1/g_{rr}(r)]^{1/2}. \tag{5-5}$$

We next need the curvature of the line of longitude:

$$k_3 = \frac{\dfrac{d^2w}{dr^2}}{\left[1 + \left(\dfrac{dw}{dr}\right)^2\right]^{3/2}} = \frac{\dfrac{dg_{rr}}{dr}}{2g_{rr}^{3/2}\sqrt{g_{rr} - 1}}. \tag{5-6}$$

Consequently, $R_{\tilde{V}3} = 2(k_1^2 + 2k_1k_3)$ gives

$$R_V = R_{\tilde{V}3} = \frac{2}{rg_{rr}}\left(\frac{g_{rr} - 1}{r} + \frac{dg_{rr}/dr}{g_{rr}}\right),$$

and combining with Equation 5-2 we get

$$\frac{dg_{rr}}{dr} - \frac{1}{r}g_{rr} = \left(8\pi\kappa\rho r - \frac{1}{r}\right)g_{rr}^2.$$

This is a simple Bernoulli equation. Put $U = 1/g_{rr}(r)$, and then solve to get

$$g_{rr} = \left(1 - \frac{\alpha}{r} - \frac{2\kappa}{r}\int 4\pi r^2\rho(r)\,dr\right)^{-1}, \tag{5-7}$$

where α is a constant.

To prevent g_{rr} from vanishing at the origin, we put $\alpha = 0$. Consider the integral

$$M(R) = \int_0^R 4\pi r^2\rho(r)\,dr. \tag{5-8}$$

Now $4\pi r^2$ is the area of the sphere of radius r but dr is *not* arc length in the radial direction; $\sqrt{g_{rr}}\,dr$ is. Consequently, $M(r)$ is *not* the volume integral of ρ over the ball bounded by S_r^2. Nevertheless, we shall still call $M(r)$ the mass of this ball, and this terminology will be justified shortly. Defining

$$m(r) = \kappa M(r),$$

again called mass, we have

$$g_{rr}(r) = \frac{1}{1 - \dfrac{2m(r)}{r}}. \tag{5-9}$$

Suppose now that our spherical ball of mass-energy has "radius" r_0, i.e., $\rho(r) = 0$ for $r > r_0$. Define

$$m = m(r_0).$$

Then $m(r) = m$ for $r \geq r_0$, and we have, for spatial metric

$$dl^2 = \frac{dr^2}{1 - \dfrac{2m(r)}{r}} + r^2(d\theta^2 + \sin^2\theta \, d\varphi^2). \tag{5-10}$$

The Gravitational Potential and g_{00}

We now must determine g_{00}. For the present we will consider g_{00} only in the region *exterior* to the ball B_{r_0} of mass-energy; we shall consider the *interior* behavior in Chapter 11. We will now assume not only that $\rho(r) = 0$ for $r > r_0$ but that the entire stress-energy-momentum tensor (T_{ij}) vanishes for $r > r_0$, i.e., space-time is empty for $r > r_0$. The generalized Poisson equation of Chapter 3 (Equations 3-9 and 3-10)

$$\nabla^2 \sqrt{-g_{00}} = 4\pi\kappa\rho^* = 4\pi\kappa(T_1^1 + T_2^2 + T_3^3 - T_0^0) \sqrt{-g_{00}} \tag{5-11}$$

can be integrated over a ball B_r, where $r > r_0$. If we put

$$M^* = \int_{B_r} \rho^* \sqrt{g_V} \, dx, \tag{5-12}$$

then we have, by Gauss's theorem,

$$\begin{aligned} 4\pi\kappa M^* &= \int_{B_r} 4\pi\kappa\rho^* \sqrt{g_V} \, dx = \int_{B_r} \nabla^2 \sqrt{-g_{00}} \sqrt{g_V} \, dx \\ &= \int_{\partial B_r} \mathsf{grad}_V \sqrt{-g_{00}} \cdot d\mathbf{S}, \end{aligned} \tag{5-13}$$

where $\partial B_r = S_r^2$ is the 2-sphere whose area is $4\pi r^2$ and where grad_V is the gradient operator in V^3. By spherical symmetry

$$\int_{\partial B_r} \mathsf{grad}_V \sqrt{-g_{00}} \cdot d\mathbf{S} = 4\pi r^2 |\mathsf{grad}_V \sqrt{-g_{00}}|. \tag{5-14}$$

It is reasonable (see p. 53ff) to call M^* as defined in Equation 5-12 the *total mass* of the ball B_r. From Equations 5-13 and 5-14 we see

$$|\mathsf{grad}_V \sqrt{-g_{00}} \, (r)| = \frac{\kappa M^*}{r^2},$$

which (since $\sqrt{-g_{00}} = 1 - U$) is *exactly* as in Newtonian gravitation, providing we continue to think of r as being defined by $r^2 = [1/(4\pi)]$ area S_r^2 and not as distance from the origin; both definitions coincide in Euclidean space.

Since

$$|\text{grad}_V \sqrt{-g_{00}}|^2 = g^{\alpha\beta} \frac{\partial \sqrt{-g_{00}}}{\partial x^\alpha} \frac{\partial \sqrt{-g_{00}}}{\partial x^\beta} = g^{rr} \left(\frac{\partial \sqrt{-g_{00}}}{\partial r}\right)^2$$

and

$$g_{rr} = 1/g^{rr},$$

we have

$$\frac{r^2}{\sqrt{g_{rr}}} \frac{\partial \sqrt{-g_{00}}}{\partial r} = \kappa M^* \equiv m^*. \tag{5-15}$$

We now invoke an equation that will be derived in Chapter 11. Let e_r be a unit vector in the radial direction, i.e., in the direction of $\partial/\partial r$. Then the Einstein equation involving $\mathsf{T}(e_r, e_r)$ becomes, in empty space (see Eqn. 11-10),

$$0 = -\frac{1}{r^2} + \frac{1}{r^2 g_{rr}} + \frac{2}{r g_{rr}} \frac{\partial}{\partial r} \log \sqrt{-g_{00}}.$$

Combining this with Equation 5-15 and using our known

$$g_{rr}(r) = \frac{1}{1 - \dfrac{2m}{r}}$$

(because $m(r) = m$ for $r \geq r_0$), we easily get

$$m^* = m \sqrt{g_{rr}} \sqrt{-g_{00}},$$

which relates the ρ "mass" m with the ρ^* "mass" m^*.

Since $g_{rr} = \left[1 - \dfrac{2m(r_0)}{r}\right]^{-1} \to 1$ as $r \to \infty$, and we have already normalized $g_{00} \to -1$ as $r \to \infty$, we see

$$m = m^* \qquad \text{and} \qquad g_{rr} g_{00} = -1. \tag{5-16}$$

We then finally have the famous Schwarzschild exterior solution in the region $r > r_0$ exterior to the ball:

$$\boxed{ds^2 = -\left(1 - \frac{2m}{r}\right) dt^2 + \frac{dr^2}{1 - \dfrac{2m}{r}} + r^2\, d\Omega^2.} \tag{5-17}$$

The Schwarzschild Singularity

Note that the coefficients g_{00} and g_{rr} become unacceptable at $r = 2m$. For most normal bodies this "Schwarzschild radius" occurs *inside* the mass, $2m < r_0$, and so there is no contradiction with our exterior solution. (For the sun the Schwarzschild radius is of the order of a few kilometers!) It is possible, though, that a ball could be so massive that its Schwarzschild radius lies outside the body, $2m > r_0$, in which case we would have to restrict our exterior solution to this region $r > 2m > r_0$. This happens in the case of a **black hole**. It is interesting though that in this case the Schwarzschild "singularity" at $r = 2m$ is actually just a singularity due to a breakdown in the *coordinates*, just as $r = 0$ is a coordinate singularity for the Euclidean metric in the plane when polar coordinates are used, $ds^2 = dr^2 + r^2\, d\theta^2$. To see that this might be the case, look at the Riemann sectional curvatures of our spatial section V^3. These curvatures are the same as those of our embedded $\tilde{V}^3 \subset R^4$. If we use again r, θ, φ as coordinates for $\tilde{V}^3$, let e_r, e_θ, and e_φ be unit vectors along the coordinate vectors. As we have seen, these are principal directions on $\tilde{V}^3$ corresponding to principal curvatures $k_r = k_3$, $k_\theta = k_1 = k_2 = k_\varphi$. Since again R^4 is flat, we have by the Gauss equations $\mathsf{K}_{\tilde{V}}(e_r \wedge e_\theta) = \mathsf{K}_{\tilde{V}}(e_r \wedge e_\varphi) = k_1 k_3$, and $\mathsf{K}_{\tilde{V}}(e_\theta \wedge e_\varphi) = k_1 k_2$. Using the values for k_1, k_2, k_3 from Equations 5-5 and 5-6, and using $g_{rr} = (1 - 2m/r)^{-1}$, we see

$$\begin{cases} \mathsf{K}_V(e_r \wedge e_\theta) = \mathsf{K}_V(e_r \wedge e_\varphi) = -\dfrac{m}{r^3} \\[2ex] \mathsf{K}_V(e_\theta \wedge e_\varphi) = \dfrac{2m}{r^3}. \end{cases} \tag{5-18}$$

Note that these expressions are also sectional curvatures of space-time, since V^3 is totally geodesic in M^4 (see Eqn. 4-10). The remaining sectional curvatures of M^4 are easily obtained from these and the fact that all Ricci curvatures vanish in the exterior region. For example,

$$0 = \mathsf{Ric}(e_r, e_r) = \mathsf{K}(e_r \wedge e_\theta) + \mathsf{K}(e_r \wedge e_\varphi) + \mathsf{K}(e_r \wedge e_t)$$

gives $\mathsf{K}(e_t \wedge e_r) = 2m/r^3$. Similarly, for the others

$$\begin{cases} \mathsf{K}(e_t \wedge e_\theta) = \mathsf{K}(e_t \wedge e_\varphi) = -\dfrac{m}{r^3} \\ \\ \mathsf{K}(e_t \wedge e_r) = \dfrac{2m}{r^3}. \end{cases} \tag{5-19}$$

We note that all of the curvatures of Equations 5-18 and 5-19 are well-behaved at the Schwarzschild radius $r = 2m$, even though g_{rr} becomes infinite and g_{00} vanishes there. This suggests that $r = 2m$ might indeed be only a singularity of the coordinate system, but it is only a suggestion, since, for example, one can not detect the singular vertex of a flat cone merely by looking at the curvature nearby. Independently, M. D. Kruskal and G. Szekeres have shown how to introduce new coordinates into the "region" $r \leq 2m$ and how to extend the Schwarzschild exterior solution. For details in these matters and their striking applications to black holes, see Misner, Thorne, and Wheeler (1973, Chapters 31 and 33).

Concluding Remarks

(1) From Equations 5-11 and 5-12 we see that M^*, rather than M, is the natural analog of the Newtonian gravitational mass of the ball. Using $g_{rr} = (1 - 2m/r)^{-1}$, Equation 5-15 can easily be solved to yield

$$\sqrt{-g_{00}} = \frac{m^*}{m}\sqrt{1 - \frac{2m}{r}} + \left(1 - \frac{m^*}{m}\right), \tag{5-20}$$

where we have chosen the constant of integration so that $\sqrt{-g_{00}} \to 1$ as $r \to \infty$. For large r we automatically have, neglecting terms in r^{-2},

$$\sqrt{-g_{00}} \sim 1 - \frac{m^*}{r} = 1 - \frac{\kappa M^*}{r}, \tag{5-21}$$

as indeed we should expect; potential should involve ρ^* and hence m^*. On the other hand, from Equation 4-15 we expect that curvature should involve ρ, and hence m, and that is why g_{rr} involves m, not m^*. It seems that we *must* invoke the other Einstein equations, as we did, in order to conclude in this spherically symmetric case that $m = m^*$. In terms of integrals, we then have two useful expressions for the mass of a ball in a static spherically symmetric world:

$$\begin{cases} \int_0^{r_0} 4\pi r^2 \rho(r)\, dr = \int_{B_{r_0}} (T^1_1 + T^2_2 + T^3_3 - T^0_0)\sqrt{-g_{00}}\,\sqrt{g_V}\, dx \\ \rho = \mathsf{T}(\xi, \xi) = -T^0_0 \end{cases} \tag{5-22}$$

a result due to R. C. Tolman, 1930. Note that $\sqrt{-g_{00}}\,\sqrt{g_V}$ is merely $\sqrt{-g}$, where $g = \det(g_{ij})$ for the full metric, and that one must integrate over the entire ball where the stress-energy-momentum is nonzero.

(2) We should also note that while Equation 5-21 exhibits $\sqrt{-g_{00}}$ as a very close analog of the Newtonian potential, the fact that $\sqrt{-g_{00}}$ satisfies a Poisson-like equation (Eq. 5-11) is more useful. For example, let $\mathcal{U}$ be an empty spherical cavity at the origin of a static, spherically symmetric mass-energy distribution. Classically, the Newtonian potential would be constant in $\mathcal{U}$. We see this and more, as follows. Inside $\mathcal{U}$, $\sqrt{-g_{00}}$ satisfies Laplace's equation $\nabla^2 \sqrt{-g_{00}} = 0$. On the other hand, by spherical symmetry $\sqrt{-g_{00}}$ is constant on the boundary $\partial\mathcal{U}$; hence, by the generalized maximum principle of E. Hopf (see Adler, Bazin, and Schiffer, 1975) we conclude that g_{00} is constant inside $\mathcal{U}$. Furthermore,

$$g_{rr} = \frac{1}{1 - \dfrac{2m(r)}{r}} = 1 \quad \text{in} \quad \mathcal{U}$$

since $m(r) = 4\pi\kappa\int \rho r^2\, dr = 0$ inside $\mathcal{U}$. Thus, inside $\mathcal{U}$ we have

$$ds^2 = -\text{ const. } dt^2 + dr^2 + r^2\, d\Omega^2$$

that is, *the metric is flat* (Minkowskian) *inside* $\mathcal{U}$.

(3) Return now to the embedding $\tilde{V}^3$ of V^3, the spatial part of the Schwarzschild universe. As we have seen (Eqns. 5-3 and 5-9), the exterior solution satisfies

$$1 + \left(\frac{dw}{dr}\right)^2 = g_{rr} = \frac{1}{1 - \dfrac{2m}{r}}.$$

If we specify, for example, that $w = 0$ when $r = 2m$, the solution is easily

$$w^2(r) = 8m(r - 2m),$$

which exhibits the exterior spatial universe as a paraboloid of revolution in R^4, the *Flamm paraboloid.*

(4) Consider a time-dependent, spherically symmetric distribution of mass-energy as would arise, for example, in the case of a radially pulsating spherical star. In Newtonian gravitation the field outside the star would be static; measurements there could not detect the pulsation. That this is also true in general relativity is a consequence of Birkhoff's theorem: a spherically symmetric time-dependent exterior solution in the region $r > 2m$ is necessarily the static Schwarzschild solution! (For a proof of this, as well as a careful treatment of many of the topics touched only briefly here, see Hawking and Ellis, 1973.)

(5) Finally, in classical potential theory Poisson's equation $\nabla^2 U = -4\pi\kappa\rho$ indicates that U, while continuously differentiable, has second derivatives that need not be continuous. This is especially clear at the surface of a body where ρ is usually only piecewise continuous. We expect then in general relativity that $\sqrt{-g_{00}}$ will not have continuous second derivatives. Consequently, *we cannot expect the curvature tensor to be continuous at the surface of a body!* This is clear from the Einstein equations. This will be illustrated in Chapter 11, where we will join an "interior" Schwarzschild solution to the exterior one of this chapter. (For general considerations on the smoothness of the metric of general relativity, see Synge, 1960.)

6

The Classical Motion of a Continuum

Lie Derivatives, Interior Products, and the Variation of Integrals

Let us recall briefly some facts concerning the action of vector fields on tensors. (For most of the details see Warner, 1971.) In the following, if M^n is a smooth n-manifold, then M^n_p will represent the tangent space to M^n at $p \in M^n$. Associated with each smooth map $f: M^n \to V^r$ is the linear "differential" action on tangent vectors, $f_* = \dot{f}: M^n_p \to V^r_{f(p)}$; if $Y \in M^n_p$, then Y is the tangent to a parameterized curve $\mathscr{C}$ thru p and then $\dot{f}Y$ is the tangent to the image curve $f \circ \mathscr{C}$ at $f(p)$. In local coordinates the matrix for $\dot{f}$ is the usual Jacobian matrix.

Let X be a smooth vector field on M^n. The integral curves of X are obtained by integrating locally the system of ordinary differential equations $dx^i/dt = X^i(x(t))$. The local one-parameter "flow" $\{\varphi_t\}$, $\varphi_t: M^n \to M^n$, moves each $p \in M^n$ for t units along the integral curve through p (which is always possible if t is sufficiently small). Each φ_t is a diffeomorphism, and for small s, t we have $\varphi_t \circ \varphi_s = \varphi_{t+s} = \varphi_s \circ \varphi_t$, $\varphi_{-t} = \varphi_t^{-1}$, and $\varphi_0 =$ the identity map.

Let $t \to Y_{\varphi_t(p)}$ be a vector field defined along the integral curve $\varphi_t(p)$ of X thru p. The Lie derivative of Y with respect to X is defined to be the vector field along the integral curve

$$(\mathscr{L}_X Y)_p = \frac{d}{dt}\,[\dot{\varphi}_{-t} Y_{\varphi_t(p)}]_{t=0}$$

$$= \lim_{t\to 0} \frac{\dot{\varphi}_{-t} Y_{\varphi_t(p)} - Y_p}{t} = \lim_{t\to 0} \frac{Y_{\varphi_t(p)} - \dot{\varphi}_t Y_p}{t}.$$

Y is said to be *invariant* under X (or φ_t) if, for each t, $\dot{\varphi}_t Y_p = Y_{\varphi_t(p)}$, and this is equivalent to $\mathscr{L}_X Y = 0$. An invariant Y satisfies *Jacobi's equation*

$$\frac{dY^i}{dt} = \frac{\partial X^i}{\partial x^j}\, Y^j$$

along the integral curve. If M^n is a (pseudo) Riemannian manifold we can compute the *intrinsic* or *covariant derivative* of this invariant Y to be

$$\frac{\nabla Y^i}{dt} = \frac{dY^i}{dt} + \Gamma^i{}_{jk} Y^k \frac{dx^j}{dt}$$

$$= \left(\frac{\partial X^i}{\partial x^k} + \Gamma^i{}_{jk} X^j\right) Y^k = X^i{}_{;k} Y^k.$$

Thus we have

$$\frac{\nabla Y}{dt} = \nabla_Y X, \tag{6-1}$$

which will also be called Jacobi's equation.

When Y itself is a vector field we have $\mathscr{L}_X Y = [X, Y]$, where $[X, Y]$ is the vector field whose differential operator is

$$[X, Y]_p(f) = X_p(Yf) - Y_p(Xf).$$

$[X, Y]$ has components

$$[X, Y]^i = X^j \frac{\partial Y^i}{\partial x^j} - Y^j \frac{\partial X^i}{\partial x^j},$$

which can be written, in a (pseudo) Riemannian M^n (see p. 36ff),

$$[X, Y]^i = X^j Y^i_{;j} - Y^j X^i_{;j}$$

or

$$[X, Y] = \nabla_X Y - \nabla_Y X.$$

For each smooth map $f: M^n \to V^r$, the *pull back* f^* maps the covariant k-tensors at $f(p)$ into covariant k-tensors at p, as follows. A covariant k-tensor is a multilinear functional on k-tuples of vectors. Let then $Y_1, \ldots, Y_k$ be vectors at $p \in M^n$ and let β^k be a k-tensor at $f(p)$. Define the pull-back by

$$(f^*\beta^k)(Y_1, \ldots, Y_k) \equiv \beta^k(\dot{f}Y_1, \ldots, \dot{f}Y_k).$$

If now α^k is a covariant k-tensor on M, define the Lie derivative $\mathscr{L}_X\alpha^k$ by

$$(\mathscr{L}_X\alpha^k) = \left[\frac{d}{dt}\, \varphi_t^* \alpha_{\varphi_t p}\right]_{t=0}$$

$$= \lim_{t\to 0} \frac{\varphi_t^* \alpha_{\varphi_t(p)} - \alpha_p}{t}.$$

In terms of its values on vectors $Y_1, \ldots, Y_k$ at p

$$(\mathscr{L}_X\alpha^k)(Y_1, \ldots, Y_k) = \left[\frac{d}{dt}\, \alpha^k(\dot{\varphi}_t Y_1, \ldots, \dot{\varphi}_t Y_k)\right]_{t=0}.$$

In particular, if the fields $Y_1, \ldots, Y_k$ are invariant under X along an integral curve, then

$$(\mathscr{L}_X\alpha^k)(Y_1, \ldots, Y_k) = \left[\frac{d}{dt}\, \alpha^k_{\varphi_t(p)}(Y_1, \ldots, Y_k)\right]_{t=0}. \quad (6\text{-}2)$$

We shall also say that α^k is invariant under X if, for all t, $\varphi_t^* \alpha^k_{\varphi_t(p)} = \alpha^k_p$, that is, if $\mathscr{L}_X\alpha^k = 0$.

If α^k is a covariant k-tensor and Y is a vector, the *interior product* (*contraction*) $i_Y\alpha$ is to be the covariant $(k-1)$ tensor defined by

$$\begin{cases} i_Y\alpha^0 = 0, \text{ if } \alpha \text{ is a 0-form, i.e., function;} \\ i_Y\alpha^1 = \alpha^1(Y), \text{ if } \alpha^1 \text{ is a 1-form (thus } i_Y\, df = Y(f)); \\ i_Y\alpha^k(Y_2, \ldots, Y_k) = \alpha^k(Y, Y_2, \ldots, Y_k) \quad \text{for } k > 1. \end{cases} \quad (6\text{-}3)$$

If α^k is an (*exterior*) k-form, that is, if α^k is a *skew symmetric* covariant k-tensor, α has a local expression

$$\alpha^k = \sum_{i_1<\cdots<i_k} a_{i_1\cdots i_k}(x)\, dx^{i_1} \wedge \cdots \wedge dx^{i_k}$$
$$= a_{i_1<\cdots<i_k}\, dx^{i_1} \wedge \cdots \wedge dx^{i_k}$$

In this expression

$$a_{i_1\cdots i_k} = \alpha^k \left(\frac{\partial}{\partial x^{i_1}}, \ldots, \frac{\partial}{\partial x^{i_k}}\right)$$

and $i_1 < \cdots < i_k$ indicates that in the implied summation *the indexes must be in increasing order*. Interior product is an antiderivation on forms, that is,

$$i_Y(\alpha^k \wedge \beta^l) = (i_Y\alpha^k) \wedge \beta^l + (-1)^k \alpha^k \wedge (i_Y\beta^l). \tag{6-4}$$

Furthermore, interior product is a form of contraction

$$i_Y\alpha^k = Y^j a_{ji_2<\cdots<i_k}\, dx^{i_2} \wedge \cdots \wedge dx^{i_k}. \tag{6-5}$$

The Lie derivative of a k-form has an especially useful expression, the *Cartan formula*:

$$\mathcal{L}_X\alpha = i_X\, d\alpha + d i_X\, \alpha. \tag{6-6}$$

A form ω^n of maximal degree, locally $\omega^n = \rho(x)\, dx^1 \wedge \cdots \wedge dx^n$, will be called a "volume form" if ρ is nonvanishing.

The *divergence* of a vector field X with respect to a volume form ω^n is defined to be that function div X such that

$$\mathcal{L}_X\omega^n = (\operatorname{div} X)\omega^n \tag{6-7}$$

Using Equations 6-6 and 6-4,

$$\begin{aligned}
&\mathcal{L}_X(\rho\, dx^1 \wedge \cdots \wedge dx^n) \\
&\quad = d[\rho i_X (dx^1 \wedge \cdots \wedge dx^n)] \\
&\quad = d[\rho(X(x^1)\, dx^2 \wedge \cdots \wedge dx^n - X(x^2)\, dx^1 \wedge dx^3 \wedge \cdots \wedge dx^n + \cdots)] \\
&\quad = d[\rho(X^1\, dx^2 \wedge \cdots \wedge dx^n - X^2\, dx^1 \wedge dx^3 \wedge \cdots \wedge dx^n + \cdots)] \\
&\quad = \frac{1}{\rho}\frac{\partial}{\partial x^i}(\rho X^i)\omega^n,
\end{aligned}$$

that is,

$$\operatorname{div} X = \frac{1}{\rho}\frac{\partial}{\partial x^i}(\rho X^i). \tag{6-8}$$

When M^n is (pseudo) Riemannian, there is a distinguished volume form; in local coordinates $\omega^n = \sqrt{|g|}\, dx^1 \wedge \cdots \wedge dx^n$, where $g = \det(g_{ij})$. One technicality: volume forms are not "true" forms but "twisted" forms. We shall discuss this in Chapter 9.

Finally, consider the "first variation" of the integral of a form. Let C_k be a k-chain on M^n (i.e., a domain of integration for k-forms) and consider the effect of the flow $\{\varphi_t\}$ generated by a vector field X. Put $C_k(t) = \varphi_t \circ C_k$ and let α^k be a k-form. Then

$$\begin{aligned}\frac{d}{dt}\int_{C_k(t)} \alpha^k &= \frac{d}{dt}\int_{\varphi_t(C_k)} \alpha^k = \frac{d}{dt}\int_{C_k} \varphi_t^* \alpha^k \\ &= \int_{C_k} \lim_{h\to 0} \frac{\varphi_{t+h}^*\alpha - \varphi_t^*\alpha}{h} \\ &= \int_{C_k} \varphi_t^* \lim_{h\to 0} \frac{\varphi_h^*\alpha - \alpha}{h} = \int_{C_k} \varphi_t^* \mathscr{L}_X \alpha.\end{aligned}$$

Thus

$$\frac{d}{dt}\int_{C_k(t)} \alpha^k = \int_{C_k(t)} \mathscr{L}_X \alpha, \tag{6-9}$$

and in particular

$$\left(\frac{d}{dt}\int_{C_k(t)} \alpha^k\right)_{t=0} = \int_{C_k} \mathscr{L}_X \alpha. \tag{6-10}$$

When C_n is an n-chain, and $\int_{C_n} \omega^n$ its volume, then

$$\begin{aligned}\frac{d}{dt}\int_{C_n(t)} \omega^n &= \int_{C_n(t)} \mathscr{L}_X \omega^n \\ &= \int_{C_n(t)} (\operatorname{div} X)\omega^n = \int_{\partial C_n(t)} i_X \omega^n,\end{aligned} \tag{6-11}$$

Equation 6-11 is a general form of the divergence theorem, following from $\mathscr{L}_X \omega^n = d i_X \omega^n$ and Stokes's theorem. If M^n is Riemannian, with volume form $\omega = \sqrt{g}\, dx^1 \wedge \cdots \wedge dx^n$, and if C_n is a compact domain with piecewise regular boundary ∂C_n, and if N is a unit normal to the regular part of ∂C_n that points out of C_n, then

$$\omega^{n-1} \equiv i_N \omega^n$$

is the $(n - 1)$ volume form for ∂C_n. If we write, on ∂C_n,

$$X = \langle X, N \rangle N + T,$$

where T is tangent to ∂C_n, then, since $i_T \omega^n$ is zero when restricted to ∂C_n, we have

$$\int_{C_n(t)} i_X \omega^n = \int_{\partial C_n(t)} \langle X, N \rangle i_N \omega^n = \int_{\partial C_n(t)} \langle X, N \rangle \omega^{n-1} \tag{6-12}$$

and one recovers the usual form of the divergence theorem

$$\int_{C_n} \operatorname{div} X \; \omega^n = \int_{\partial C_n} \langle X, N \rangle \omega^{n-1}. \tag{6-13}$$

While this is usually stated for *orientable* manifolds, it actually holds also for nonorientable ones, since ω^n and ω^{n-1} are actually twisted forms or forms of "odd kind" in the sense of de Rham (these forms will be discussed briefly in Chapter 9).

We now adapt the above material to the classical case of a time-dependent vector field $\mathbf{v}(t, \mathbf{x})$ in ordinary 3-space R^3. This is most easily handled by introducing $R \times R^3$ and the new vector field (not to be confused with similar fields introduced in relativity) on $R \times R^3$:

$$u = \frac{\partial}{\partial t} + \mathbf{v}(t, \mathbf{x}).$$

A time-dependent k-form on R^3 is really a k-form on $R \times R^3$ having no terms involving dt. If $C_k(0)$ is a k-chain on R^3 at $t = 0$, we can consider the future images $C_k(t)$ of $C_k(0)$ under the time-dependent flow generated by the time-dependent field $\mathbf{v}(t, \mathbf{x})$; they are most easily pictured as the projections into R^3 of corresponding chains $C_k(t)$ in $R \times R^3$,

$$C_k(t) = \varphi_t \circ C_k(0),$$

where $\{\varphi_t\}$ is the flow on $R \times R^3$ generated by the vector field u (Figure 6-1).

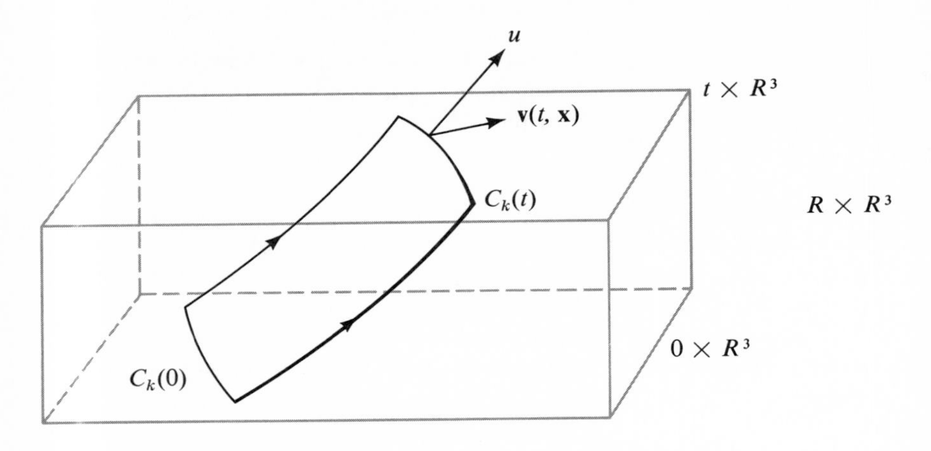

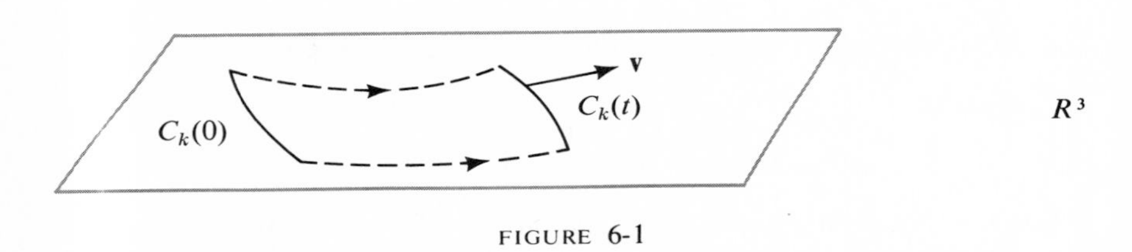

FIGURE 6-1

We now introduce the following notation adapted to $R \times R^3$. If $\alpha^k = \alpha^k(t, \mathbf{x})$ is a time-dependent form on R^3, let $\dfrac{\partial \alpha^k}{\partial t}$ be the time-dependent k-form whose coefficients are the time derivatives of those of α^k. Let

$$\mathbf{d} = dx^1 \wedge \frac{\partial}{\partial x^1} + dx^2 \wedge \frac{\partial}{\partial x^2} + dx^3 \wedge \frac{\partial}{\partial x^3}$$

and

$$d = \mathbf{d} + dt \wedge \frac{\partial}{\partial t}$$

be the exterior differential operators on $t \times R^3$ and $R \times R^3$, respectively. Then from Equation 6-9

$$\frac{d}{dt} \int_{C_k(t)} \alpha^k = \int_{C_k(t)} \mathscr{L}_{\partial/\partial t + \mathbf{v}}\, \alpha^k = \int_{C_k(t)} \frac{\partial \alpha^k}{\partial t} + \mathscr{L}_{\mathbf{v}} \alpha^k,$$

and a computation using Equation 6-6 yields

$$\frac{d}{dt}\int_{C_k(t)} \alpha^k = \int_{C_k(t)} \frac{\partial \alpha^k}{\partial t} + i_{\mathbf{v}}\mathbf{d}\alpha^k + \mathbf{d}i_{\mathbf{v}}\alpha^k. \tag{6-14}$$

This formula contains all the classical expressions of vector analysis for the differentiation of line, surface, and volume integrals. This can be seen from an easily constructed "dictionary" relating vector analysis to exterior forms. For simplicity we consider Cartesian coordinates x, y, z for R^3 and we shall revert to using d for exterior derivative.

Dictionary

Let

$\omega^3 = dx \wedge dy \wedge dx$, the volume form

1-form $\alpha^1 =$ the covariant expression for a vector **A**

1-form $\gamma^1 =$ the covariant expression for a vector **C**

2-form β^2, identified with the unique vector **B** such that

$$\beta^2 = i_{\mathbf{B}}\omega^3$$

Then

$\alpha^1 \wedge \gamma^1 = i_{\mathbf{A}\times\mathbf{C}}\omega^3$ is identified with $\mathbf{A} \times \mathbf{C}$

$\alpha^1 \wedge \beta^2 = (\mathbf{A} \cdot \mathbf{B})\omega^3$ is identified with $\mathbf{A} \cdot \mathbf{B}$

$i_{\mathbf{C}}\alpha^1 = \mathbf{C} \cdot \mathbf{A}$

$i_{\mathbf{C}}\beta^2 =$ the covariant expression for $\mathbf{B} \times \mathbf{C}$

$d\alpha^1 = i_{\text{curl}\,\mathbf{A}}\omega^3$ is identified with curl **A**

$d\beta^2 = \text{div}\,\mathbf{B}\,\omega^3$ is identified with div **B**.

Using these definitions, Equation 6-14 yields the classical expressions

$$\begin{cases} \dfrac{d}{dt}\displaystyle\int_{C_1} \mathbf{A}\cdot d\mathbf{r} = \int_{C_1}\left[\frac{\partial \mathbf{A}}{\partial t} - \mathbf{v}\times \text{curl}\,\mathbf{A} + \text{grad}(\mathbf{v}\cdot\mathbf{A})\right]\cdot d\mathbf{r} \\ \dfrac{d}{dt}\displaystyle\int_{C_2} \mathbf{B}\cdot\mathbf{N}\,dA = \int_{C_2}\left[\frac{\partial \mathbf{B}}{\partial t} + (\text{div}\,\mathbf{B})\mathbf{v} - \text{curl}(\mathbf{v}\times\mathbf{B})\right]\cdot\mathbf{N}\,dA \\ \dfrac{d}{dt}\displaystyle\int_{C_3} \rho\omega^3 = \int_{C_3}\left[\frac{\partial\rho}{\partial t} + \text{div}(\rho\mathbf{v})\right]\omega^3. \end{cases} \qquad (6\text{-}15)$$

Note explicitly that Equation 6-6 yields convenient vector-analytic expressions for $\mathscr{L}_{\mathbf{A}}(\gamma^1)$ and $\mathscr{L}_{\mathbf{A}}(\beta^2)$ in the time-independent case.

The Cauchy Stress Tensor in Classical Mechanics

Let R^3 be ordinary 3-space with a given right-handed Cartesian coordinate system. We shall write all vectors in purely covariant form. Consider a nonrelativistic time-dependent fluid flow in R^3, with density $\rho = \rho(t, \mathbf{x})$ and velocity vector $\mathbf{v} = \mathbf{v}(t, \mathbf{x})$. If we follow a region B_0 under the flow to B_t, we have for the mass of the fluid in B_t

$$m(t) = \int_{B_t} \rho\omega^3,$$

and by the third equation of Equation 6-15

$$\frac{dm}{dt} = \int_{B_t}\left[\frac{\partial\rho}{\partial t} + \text{div}(\rho\mathbf{v})\right]\omega^3.$$

Conservation of mass (which holds classically, but not in special relativity) states $dm/dt = 0$, or, from Equation 6-9,

$$\mathscr{L}_u(\rho\omega^3) = 0 = \frac{\partial\rho}{\partial t} + \text{div}(\rho\mathbf{v}). \qquad (6\text{-}16)$$

If now $f = f(t, \mathbf{x})$ is any smooth function, we have, from Equation 6-16,

$$\frac{d}{dt}\int_{B_t} f\rho\omega^3 = \int_{B_t} \mathscr{L}_u(f\rho\omega^3) = \int_{B_t} \frac{df}{dt}\,\rho\omega^3, \qquad (6\text{-}17)$$

where $df/dt = \partial f/\partial t + \mathbf{v} \cdot \mathsf{grad} f = u(f)$. The linear momentum of the fluid in B_t is by definition

$$\mathbf{P} = \int_{B_t} \mathbf{v}\rho\omega^3,$$

which is merely an abbreviation for the three scalar integrals

$$P_\alpha = \int_{B_t} v_\alpha \rho\omega^3.$$

If $\mathbf{F}$ is the total force acting on B_t, Newton's second law of motion (as formulated for a continuum by Euler) states that $\mathbf{F} = d\mathbf{P}/dt$, or (from Eqn. 6-17)

$$F_\alpha = \frac{dP_\alpha}{dt} = \frac{d}{dt}\int_{B_t} v_\alpha \rho\omega^3 = \int_{B_t} \frac{dv_\alpha}{dt}\,\rho\omega^3,$$

or

$$\mathbf{F} = \int_{B_t} \frac{d\mathbf{v}}{dt}\,\rho\omega^3.$$

We assume, following Cauchy, that the total force on B_t consists of two contributions. First, there are *body* forces (for example, the gravitational pull of the earth on the body):

$$\mathbf{F}_b = \int_{B_t} \mathbf{b}\omega^3,$$

where $\mathbf{b}$ is the force density per unit volume. Second, there are contact or surface forces, or *stresses* (Figure 6-2):

$$\mathbf{F}_s = \int_{\partial B_t} \mathsf{S}(\mathbf{N})\, dA\,.$$

Here $\mathbf{N}$ is the outward-pointing unit normal to ∂B_t, $dA = i_{\mathbf{N}}\omega^3$ is the area 2-form, and $\mathsf{S}(\mathbf{N})$ is the density of force (per unit surface area) that the portion of the fluid on the side of dA to which $\mathbf{N}$ points exerts on the portion on the other side of dA. For example, if B_t is a region inside a perfect fluid then $\mathsf{S}(\mathbf{N}) = -p\mathbf{N}$ where p is the fluid *pressure*.

Euler's equation $\mathbf{F} = d\mathbf{P}/dt$ becomes

$$\frac{d}{dt}\int_{B_t} \mathbf{v}\rho\omega^3 = \int_{B_t} \frac{d\mathbf{v}}{dt}\,\rho\omega^3 = \int_{B_t} \mathbf{b}\omega^3 + \int_{\partial B_t} \mathsf{S}(\mathbf{N})\, dA\,. \tag{6-18}$$

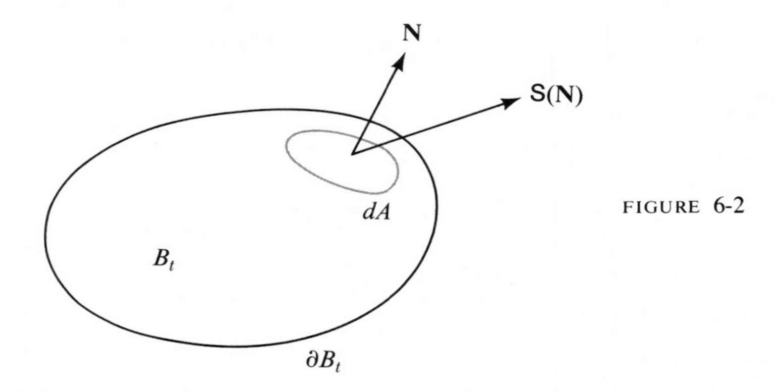

FIGURE 6-2

So far the stress S is defined only for unit normals; it is natural to extend the definition so as to make S positively homogeneous, $\mathsf{S}(\lambda\mathbf{N}) = \lambda\mathsf{S}(\mathbf{N})$ for $\lambda \geq 0$ (one can then write expressions such as $\mathsf{S}(\mathbf{N}\,dA)$). Consider a surface Σ that divides a small body B into two parts, B_+ and B_-. Let $\mathbf{N}_+$ (resp. $\mathbf{N}_-$) be the outward-pointing unit normal to B_+ (resp. B_-) (Figure 6-3). Applying Equation 6-18 to B, B_+, and B_- shows that $\int_\Sigma [\mathsf{S}(\mathbf{N}_+) + \mathsf{S}(\mathbf{N}_-)]\,dA = 0$, and since Σ is arbitrary we conclude $\mathsf{S}(-\mathbf{N}) = -\mathsf{S}(\mathbf{N})$, and so S is a *homogeneous* function of $\mathbf{N}$.

We now show that S is a *linear* function of $\mathbf{N}$. For this, we consider a small tetrahedron B (Figure 6-4). Let the unit normal $\mathbf{N}$ to the "roof" have components N_x, N_y, N_z and let the roof have area A. Then the sides perpendicular to the axes have areas $A_x = N_xA$, $A_y = N_yA$, and $A_z = N_zA$. Introduce the notation $\{U\}$ to indicate either a volume average or else a surface area average over a given surface for the quantity U. For example,

$$\frac{d}{dt}\int_{B_t} \mathbf{v}\rho\omega^3 = \int_{B_t} \frac{d\mathbf{v}}{dt}\,\rho\omega^3 = \left\{\frac{d\mathbf{v}}{dt}\,\rho\right\}\cdot \mathsf{vol}(B_t).$$

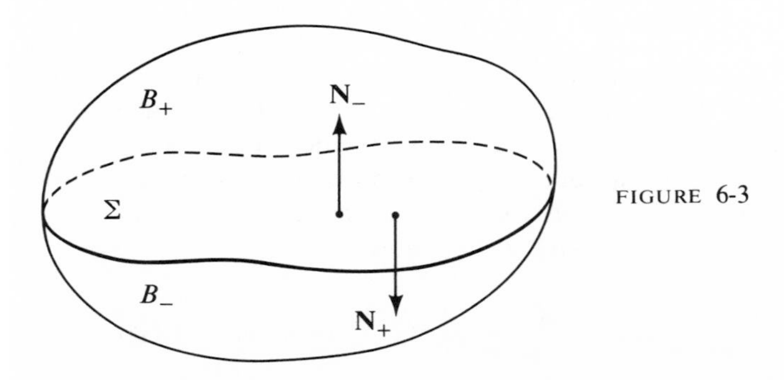

FIGURE 6-3

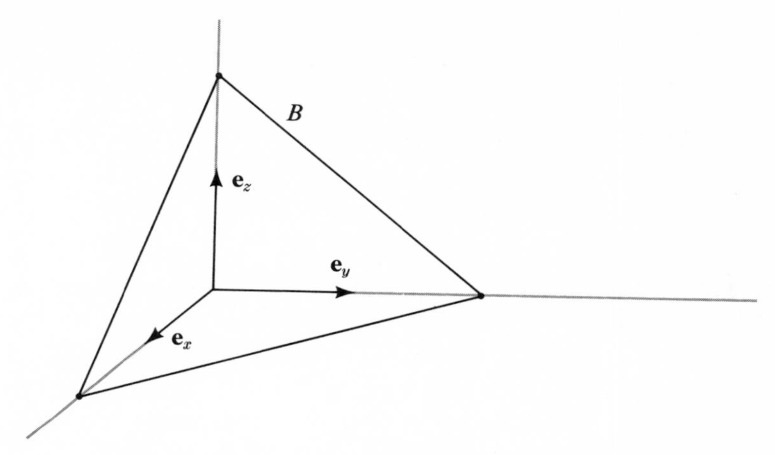

FIGURE 6-4

Then, since $\mathbf{e}_x$, $\mathbf{e}_y$, $\mathbf{e}_z$ are *inward*-pointing normals to the faces, we have

$$\begin{aligned}\int_{\partial B_t} \mathsf{S}(\mathbf{N})\, dA &= A \cdot \{\mathsf{S}(\mathbf{N})\} - A_x\{\mathsf{S}(\mathbf{e}_x)\} - A_y\{\mathsf{S}(\mathbf{e}_y)\} - A_z\{\mathsf{S}(\mathbf{e}_z)\} \\ &= A \cdot [\{\mathsf{S}(\mathbf{N})\} - (N_x\{\mathsf{S}(\mathbf{e}_x)\} + N_y\{\mathsf{S}(\mathbf{e}_y)\} + N_z\{\mathsf{S}(\mathbf{e}_z)\})].\end{aligned}$$

Thus, from Equation 6-18,

$$\left[\left\{\frac{d\mathbf{v}}{dt}\,\rho\right\} - \{\mathbf{b}\}\right] \cdot \frac{\mathrm{vol}(B)}{A} = \{\mathsf{S}(\mathbf{N})\} - (N_x\{\mathsf{S}(\mathbf{e}_x)\} + N_y\{\mathsf{S}(\mathbf{e}_y)\} + N_z\{\mathsf{S}(\mathbf{e}_z)\}).$$

Now let the tetrahedron shrink to a point while retaining the same proportions. From $\mathrm{vol}(B)/A \to 0$ we get Cauchy's Principal Lemma: $\mathsf{S}(\mathbf{N})$ is a linear function of $\mathbf{N}$:

$$\mathsf{S}(\mathbf{N}) = N_x\mathsf{S}(\mathbf{e}_x) + N_y\mathsf{S}(\mathbf{e}_y) + N_z\mathsf{S}(\mathbf{e}_z).$$

The linear transformation S is called the *Cauchy stress tensor*:

$$\mathsf{S}(\mathbf{n})_\alpha = \sum_\beta S_{\alpha\beta} n_\beta.$$

$(S_{\alpha\beta})$ are its components with respect to the usual basis $\mathbf{e}_x$, $\mathbf{e}_y$, $\mathbf{e}_z$. The αth component of the stress force $\mathbf{F}_s$ is then

$$\int_{\partial B_t} \sum_\beta S_{\alpha\beta} N_\beta \, dA = \int_{B_t} \sum_\beta \frac{\partial S_{\alpha\beta}}{\partial x^\beta} \omega^3.$$

Thus if one defines the *vector* Div S by

$$(\text{Div } \mathsf{S})_\alpha = \sum_\beta \frac{\partial S_{\alpha\beta}}{\partial x^\beta}$$

we can apply Euler's equation (Eqn. 6-18) to arbitrarily small regions B_t to conclude Cauchy's First Fundamental Theorem:

$$\begin{cases} \rho \dfrac{d\mathbf{v}}{dt} = \mathbf{b} + \text{Div } \mathsf{S} \\[2ex] \dfrac{dv_\alpha}{dt} = \dfrac{\partial v_\alpha}{\partial t} + \mathbf{v} \cdot \text{grad } v_\alpha. \end{cases} \tag{6-19}$$

In the special case of a perfect fluid the stress is isotropic, $S_{\alpha\beta} = -p\delta_{\alpha\beta}$, and Equation 6-19 then yields the classical

$$\rho \frac{d\mathbf{v}}{dt} = \mathbf{b} - \text{grad}\, p. \tag{6-20}$$

Finally, consider the angular momentum $\mathbf{H}$ of a body B about a given origin O:

$$\mathbf{H} = \int_{B_t} \mathbf{r} \times \mathbf{v}\, \rho\omega^3,$$

where $\mathbf{r}$ is the position vector from O. Euler's rotational equations state that $d\mathbf{H}/dt$ must equal the impressed torques

$$\frac{d\mathbf{H}}{dt} = \int_{B_t} \mathbf{r} \times \frac{d\mathbf{v}}{dt} \rho\omega^3 = \int_{B_t} \mathbf{r} \times \mathbf{b}\omega^3 + \int_{\partial B_t} \mathbf{r} \times \mathsf{S}(\mathbf{N})\, dA.$$

Using Equation 6-19 we get

$$\int_{B_t} \mathbf{r} \times \text{Div } \mathsf{S}\, \omega^3 = \int_{\partial B_t} \mathbf{r} \times \mathsf{S}(\mathbf{N})\, dA.$$

Now from our dictionary (p. 63), the cross product $A \times B$ can be identified with the skew tensor whose $\alpha\beta$ component is $A_\alpha B_\beta - A_\beta B_\alpha$. Thus the above equation yields

$$\int_{B_t}\left(x_\alpha \sum_\gamma \frac{\partial S_{\beta\gamma}}{\partial x^\gamma} - x_\beta \sum_\gamma \frac{\partial S_{\alpha\gamma}}{\partial x^\gamma}\right)\omega^3 = \int_{\partial B_t} [x_\alpha \mathsf{S}(N)_\beta - x_\beta \mathsf{S}(N)_\alpha]\, dA\,.$$

But $\mathsf{S}(N)_\beta = \sum_\gamma S_{\beta\gamma}N_\gamma$, and the divergence theorem yields then for the right-hand side

$$\int_{\partial B_t} \sum_\gamma (x_\alpha S_{\beta\gamma}N_\gamma - x_\beta S_{\alpha\gamma}N_\gamma)\, dA$$

$$= \int_{B_t} \sum_\gamma \frac{\partial}{\partial x^\gamma}(x_\alpha S_{\beta\gamma} - x_\beta S_{\alpha\gamma})\omega^3$$

$$= \int_{B_t} (S_{\beta\alpha} - S_{\alpha\beta})\omega^3 + \int_{B_t} \sum_\gamma \left(x_\alpha \frac{\partial S_{\beta\gamma}}{\partial x^\gamma} - x_\beta \frac{\partial S_{\alpha\gamma}}{\partial x^\gamma}\right)\omega^3.$$

From these two equations we conclude Cauchy's Second Fundamental Theorem: the stress tensor is symmetric,

$$S_{\beta\alpha} = S_{\alpha\beta}.$$

The Stress-Energy-Momentum Tensor

The laws of conservation of mass (Eqn. 6-16) and balance of linear momentum (Eqn. 6-19) can be combined into one elegant law in four-dimensional nonrelativistic space-time $R \times R^3$. For future purposes we will now write all tensors in purely contravariant form. First, extend the 3×3 stress tensor $(S^{\alpha\beta})$ to a 4×4 matrix (S^{ij}) by

$$S^{\alpha\beta} = \text{the stress tensor for } \alpha, \beta, = 1, 2, 3$$

$$S^{0j} = S^{j0} = 0 \text{ for } j = 0, 1, 2, 3.$$

Let $u = \begin{pmatrix} 1 \\ \mathbf{v} \end{pmatrix} = \partial/\partial t + \mathbf{v}$ be the classical velocity 4-vector, and define the "matter tensor" $\mathsf{M} = \rho u \otimes u$, i.e.,

$$M^{ij} = \rho u^i u^j,$$

where ρ is again the 3-volume density of matter. Finally, define the stress-energy-momentum-tensor by

$$T^{ij} = M^{ij} - S^{ij} = \rho u^i u^j - S^{ij}. \tag{6-21}$$

It is a *symmetric* tensor. Define now the body 4-force volume density by

$$b = \begin{pmatrix} 0 \\ \mathbf{b} \end{pmatrix}.$$

A simple calculation then shows that Equations 6-16 and 6-19 yield

$$\begin{cases} \displaystyle\sum_j \frac{\partial T^{0j}}{\partial x^j} = \frac{\partial \rho}{\partial t} + \sum_\beta \frac{\partial}{\partial x^\beta}(\rho v^\beta) = 0 \\ \displaystyle\sum_j \frac{\partial T^{\alpha j}}{\partial x^j} = \rho \frac{dv^\alpha}{dt} - \sum_\beta \frac{\partial S^{\alpha\beta}}{\partial x^\beta} = b^\alpha \end{cases}$$

and so Equations 6-16 and 6-19 can be expressed as the divergence

$$\sum_{j=0}^{3} \frac{\partial T^{ij}}{\partial x^j} = b^i, \tag{6-22}$$

in which the right side represents the volume density of "external" force. These equations are the classical *equations of motion*, written in four-dimensional language.

7

The Relativistic Equations of Motion

Fermi Transport and the Relative Velocity Vector

The *unit* tangent vectors along a family of world lines arise from using proper time as parameter. For many purposes, however, it is convenient to introduce a new parameterization. Consider two nearby world lines $\mathscr{C}_0$ and $\mathscr{C}$, each parameterized by its proper time τ, and suppose that a geodesic segment γ_0 that leaves $\mathscr{C}_0$ orthogonally at $\mathscr{C}_0(0)$ strikes $\mathscr{C}$ at $\mathscr{C}(0)$. Follow each world line near $\mathscr{C}_0$ for proper time τ. Then the geodesic segment γ_0 will be sent into a new segment γ_τ (which in general will not be a geodesic) that joins $\mathscr{C}_0(\tau)$ to $\mathscr{C}(\tau)$, but γ_τ will not, in general, be orthogonal to $\mathscr{C}_0$. Thus while γ_0 will consist of events that are, in the language of special relativity, approximately simultaneous with $\mathscr{C}_0(0)$, the events of γ_τ will not be approximately simultaneous with $\mathscr{C}_0(\tau)$ (Figure 7-1). We can remedy this situation by reparameterizing the world lines near $\mathscr{C}_0$ as follows. Let $e_1(\tau)$, $e_2(\tau)$, $e_3(\tau)$ be independent vector fields defined along $\mathscr{C}_0$ that span, for each τ, the vector space $u_\tau^\perp$ orthogonal to

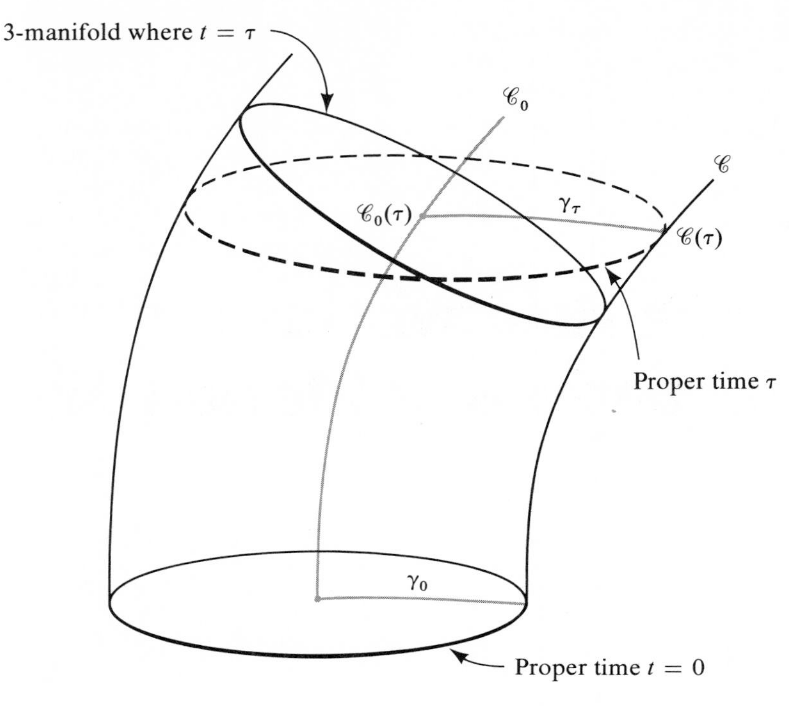

FIGURE 7-1

$\mathscr{C}_0$ at $\mathscr{C}_0(\tau)$. Locally, we can introduce coordinates in a tubular neighborhood of $\mathscr{C}_0$ by

$$(t, x^1, x^2, x^3) \to \exp_{\mathscr{C}_0(t)}[x^1 e_1(t) + x^2 e_2(t) + x^3 e_3(t)], \tag{7-1}$$

where, as usual, $\exp_p X$ is the point at distance $\|X\|$ from p on the geodesic whose tangent at p is X. Then t is a well-defined function in this tubular neighborhood, t coincides with proper time τ on $\mathscr{C}_0$, and the three-dimensional submanifolds of this neighborhood defined by putting $t =$ constant are orthogonal to $\mathscr{C}_0$. A world line $\mathscr{C}$ that is sufficiently close to $\mathscr{C}_0$ can be parameterized by t. The resulting tangent vectors

$$\tilde{u} = \frac{d\mathscr{C}}{dt} = \frac{d\tau}{dt}\frac{d\mathscr{C}}{d\tau} = \varphi u, \tag{7-2}$$

where $\varphi = d\tau/dt$, generate then a flow $\{\psi_t\}$ near $\mathscr{C}_0$ such that $\dot{\psi}_t : u_0^\perp \to u_t^\perp$, since ψ_t is merely translation by t.

We shall call a vector field $t \to X(t)$ defined along $\mathscr{C}_0$ a *relative position vector* (for a "nearby" world line) if $X(t) \in u_t^\perp$ and X is invariant under the flow $\{\psi_t\}$, i.e., $\dot{\psi}_t X(0) = X(t)$.

We wish now to define a relative (spatial) velocity vector along $\mathscr{C}_0$; the immediate candidate $\nabla X/dt$ (covariant derivative) is not suitable, since this vector need not lie in $u^\perp$. We proceed as follows, replacing the notion of parallel translation (which leads to $\nabla X/dt$) by one of *Fermi translation*. First, let $t \to X(t)$ be any vector field defined along $\mathscr{C}_0$ such that $X(t) \in u_t^\perp$ for all t. We shall say that X is Fermi-translated along $\mathscr{C}_0$ if $\nabla X/d\tau$ has no component in $u_t^\perp$, i.e., $\nabla X/d\tau = \lambda \cdot u$ for some function λ. Since $X(t) \in u_t^\perp$, we have $0 = \langle X, u\rangle = d/d\tau\langle X, u\rangle = \left\langle \frac{\nabla X}{d\tau}, u\right\rangle + \langle X, \nabla_u u\rangle = -\lambda + \langle X, \nabla_u u\rangle$. Thus X is Fermi-translated along $\mathscr{C}_0$ if X satisfies the linear differential equation $\mathsf{F}_u X = 0$, where

$$\mathsf{F}_u X \equiv \frac{\nabla X}{d\tau} - \langle X, \nabla_u u\rangle u. \tag{7-3}$$

Conversely, given $X_0 \in u_0^\perp$, there is a unique vector field X defined along $\mathscr{C}_0$ satisfying $\mathsf{F}_u X = 0$ and $X(0) = X_0$, and automatically $X(t) \in u_t^\perp$ for all t. If $X \in u_t^\perp$ and $Y \in u_t^\perp$ are both Fermi-translated along $\mathscr{C}_0$ then

$$\frac{d}{d\tau}\langle X, Y\rangle = \left\langle \frac{\nabla X}{d\tau}, Y\right\rangle + \left\langle X, \frac{\nabla Y}{d\tau}\right\rangle = \langle \mathsf{F}_u X, Y\rangle + \langle X, \mathsf{F}_u Y\rangle = 0.$$

Fermi translation yields isometries of the normal spaces $u^\perp$. $\mathsf{F}_u X$ as defined by Equation 7-3 is the *Fermi derivative* of X. (I should point out that we shall only apply Fermi derivatives to fields in $u^\perp$; it is sometimes useful to extend the notion of the Fermi derivative to all vector fields along $\mathscr{C}_0$ by defining $\mathsf{F}_u u = 0$ and insisting that F_u be a derivation.)

Note that $\mathsf{F}_u = \nabla/d\tau$ when $\mathscr{C}_0$ is a geodesic, i.e., $\nabla_u u = 0$.

Finally, let $t \to X(t)$ be a relative position vector along $\mathscr{C}_0$, i.e., $X(t) \in u_t^\perp$ and X is invariant under the flow $\{\psi_t\}$ generated by $\tilde{u} = \varphi u$. We define the *relative velocity vector* at the position $X(t)$ to be the vector

$$v(X) = \mathsf{F}_u X. \tag{7-4}$$

From
$$\langle \mathsf{F}_u X, u\rangle = \left\langle \frac{\nabla X}{d\tau}, u\right\rangle + \langle X, \nabla_u u\rangle$$
$$= \frac{d}{d\tau}\langle X, u\rangle - \langle X, \nabla_u u\rangle + \langle X, \nabla_u u\rangle = 0$$

we see that $v(X) \in u_t^\perp$ for all t.

Vorticity, Strain, and Expansion

Let $v(X)$ be the relative velocity vector at the position X. Since X is invariant under the flow generated by $\tilde{u} = \varphi u$, we have, from Equation 6-1, $\nabla X/dt = \nabla_X \tilde{u}$ along $\mathscr{C}_0$. Then for any field $t \to Y(t) \in u_t^\perp$ we have, since $\varphi = 1$ along $\mathscr{C}_0$,

$$\begin{aligned}\langle v(X), Y\rangle &= \langle F_u X, Y\rangle = \left\langle \frac{\nabla X}{d\tau}, Y\right\rangle = \left\langle \frac{\nabla X}{dt}, Y\right\rangle \\ &= \langle \nabla_X \tilde{u}, Y\rangle = \langle \nabla_X(\varphi u), Y\rangle \\ &= X(\varphi)\langle u, Y\rangle + \varphi\langle \nabla_X u, Y\rangle = \langle \nabla_X u, Y\rangle.\end{aligned}$$

Thus we may also write for the velocity vector

$$v(X) = \nabla_X u. \tag{7-5}$$

We now associate to this family of world lines (or flow $\{\psi_t\}$) two bilinear forms: $u_t^\perp \times u_t^\perp \to R$. For X and Y in $u_t^\perp$, put

$$\begin{cases} 2\Theta(X, Y) = \langle X, \nabla_Y u\rangle + \langle \nabla_X u, Y\rangle \\ 2\Omega(X, Y) = \langle X, \nabla_Y u\rangle - \langle \nabla_X u, Y\rangle, \end{cases} \tag{7-6}$$

which are the symmetric and antisymmetric forms associated to the form $(X, Y) \to \langle X, \nabla_Y u\rangle$. The term Θ is the *rate of strain* form, and Ω the *vorticity* or *rate of rotation* form. We shall now investigate the geometric significance of these forms.

First, note that if $\{e_0, e_1, e_2, e_3\}$ is any basis for vector fields defined near an event p, then for any vector field Z we have from the equation on page 35 (referring all vectors to the given basis)

$$\nabla_{e_i} Z = \nabla_{e_i}(Z^j e_j) = Z^j{}_{;k} e_i^k e_j = Z^j{}_{;i} e_j.$$

Choose a basis such that $e_0 = u$ and $\{e_1, e_2, e_3\}$ is a basis for $u_t^\perp$. Then at $\mathscr{C}_0(t)$ we have

$$\begin{aligned}2\Theta(e_\alpha, e_\beta) &= \langle \nabla_{e_\alpha} u, e_\beta\rangle + \langle e_\alpha, \nabla_{e_\beta} u\rangle \\ &= u^j{}_{;\alpha}\langle e_j, e_\beta\rangle + u^j{}_{;\beta}\langle e_\alpha, e_j\rangle \\ &= u^j{}_{;\alpha} g_{j\beta} + u^j{}_{;\beta} g_{\alpha j}\end{aligned}$$

with a similar computation for $\Omega(e_\alpha, e_\beta)$. Thus

$$\begin{cases} \theta_{\alpha\beta} = \Theta(e_\alpha, e_\beta) = \dfrac{1}{2}(u_{\alpha;\beta} + u_{\beta;\alpha}) \\ \\ \omega_{\alpha\beta} = \Omega(e_\alpha, e_\beta) = \dfrac{1}{2}(u_{\alpha;\beta} - u_{\beta;\alpha}). \end{cases} \tag{7-7}$$

Now specialize the basis $\{e_0, e_1, e_2, e_3\}$; let $e_0 = u$ and let $e_1(t)$, $e_2(t)$, $e_3(t)$ result from Fermi-translating an orthonormal basis $e_1(0)$, $e_2(0)$, $e_3(0)$ of $u_0^\perp$. For a relative position vector along $\mathscr{C}_0$,

$$X(t) = \sum X^\alpha(t) e_\alpha(t),$$

we have the relative velocity field

$$v(X(t)) = \mathsf{F}_u X(t) = \sum \frac{dX^\alpha}{dt}(t) e_\alpha(t).$$

Fermi-translation allows us to identify each $u_t^\perp$ with $u_0^\perp$ ($X(t)$ is then identified with $X(0)$ if $X^\alpha(t) = X^\alpha(0)$, $\alpha = 1, 2, 3$). Putting $\mathbf{e}_\alpha = e_\alpha(0)$, we have, in this single $u_0^\perp$, the position vector

$$\mathbf{x}(t) = \sum_{\alpha=1}^{3} x_\alpha(t) \mathbf{e}_\alpha \tag{7-8}$$

where $x_\alpha = X^\alpha$. We may use purely covariant placement of indexes since $\mathbf{e}_1$, $\mathbf{e}_2$, $\mathbf{e}_3$ are orthonormal. $\mathbf{x}(t)$ is then a time-dependent position vector in a single Euclidean $u_0^\perp$. The velocity field becomes a time-dependent vector field defined on all of $u_0^\perp$:

$$\mathbf{v}(\mathbf{x}(t)) = \sum_{\alpha=1}^{3} \frac{dx_\alpha}{dt} \mathbf{e}_\alpha = \sum_{\alpha=1}^{3} v_\alpha(\mathbf{x}(t)) \mathbf{e}_\alpha. \tag{7-9}$$

Since $X(t)$ is invariant under the flow ψ_t generated by $\tilde{u}$, $X(t) = \dot{\psi}_t X(0)$, we can write

$$\mathbf{x}(t) = \mathsf{A}(t)\mathbf{x}(0), \tag{7-10}$$

where $\mathsf{A}(t) = \dot{\psi}_t$. Consider the polar decomposition $\mathsf{A}(t) = \mathsf{R}(t)\mathsf{P}(t)$ of the nonsingular $\mathsf{A}(t)$. For each t, $\mathsf{R}(t)$ is a rotation of $u_0^\perp$ and $\mathsf{P}(t)$ is a positive definite transformation representing "stretches" along three mutually orthogonal eigendirections. Using a dot to represent ordinary time derivatives,

$$\dot{\mathsf{A}}(0) = \dot{\mathsf{R}}(0) + \dot{\mathsf{P}}(0) \tag{7-11}$$

since $\mathsf{R}(0) = \mathsf{I} = \mathsf{P}(0)$. The term $\dot{\mathsf{P}}(0)$ is again a symmetric transformation (called the rate of strain). On the other hand, for small t, $\mathsf{R}(t)$ has a unique expression in terms of an "infinitesimal generator" $\mathsf{S}(t)$:

$$\mathsf{R}(t) = e^{\mathsf{S}(t)} = \mathsf{I} + \mathsf{S}(t) + \frac{1}{2}\,\mathsf{S}^2(t) + \cdots$$

where $\mathsf{S}(t)$ is skew for all t. Since $\mathsf{S}(0) = 0$, $\dot{\mathsf{R}}(0) = \dot{\mathsf{S}}(0)$ is skew symmetric, and thus Equation 7-11 is merely the decomposition of $\dot{\mathsf{A}}(0)$ into skew and symmetric parts. From Equations 7-5, 7-9, and 7-10 we see $(\nabla_X u)_{\mathscr{C}_0} = \mathbf{v}(\mathbf{x}(0)) = \dot{\mathsf{A}}(0)\mathbf{x}(0)$, and so $\dot{\mathsf{A}}(0)$ is the linear transformation

$$X \xrightarrow{\dot{\mathsf{A}}(0)} \nabla_X u, \tag{7-12}$$

and consequently the matrix for $\dot{\mathsf{A}}(0)$ is $\dot{\mathsf{A}}_{\alpha\beta}(0) = \langle \nabla_{e_\beta} u,\, e_\alpha \rangle = u_{\alpha;\beta}$. Hence, from Equation 7-11,

$$\begin{cases} \dot{P}_{\alpha\beta}(0) = \dfrac{1}{2}\,\{\dot{A}_{\alpha\beta}(0) + \dot{A}_{\beta\alpha}(0)\} = \theta_{\alpha\beta} \\[2ex] \dot{R}_{\alpha\beta}(0) = \dfrac{1}{2}\,\{\dot{A}_{\alpha\beta}(0) - \dot{A}_{\beta\alpha}(0)\} = \omega_{\alpha\beta}. \end{cases} \tag{7-13}$$

These supply the geometric significance of Θ and Ω in terms of rates of stretching (strain) and rotation, all with respect to a Fermi-translated basis.

The flow is said to be *vorticity free* or *irrotational* if the form Ω vanishes. This form Ω is intimately connected with the distribution of 3-planes $u^\perp$ orthogonal to the flow. Let the 1-form ν be the covariant version of the unit velocity field u, $\nu(Y) = \langle u, Y\rangle$ for all Y. In local coordinates $\nu = u_i dx^i$. Then a necessary and sufficient condition that the distribution $u^\perp$ be *integrable* (i.e., that one can construct 3-manifolds of M^4 that are always orthogonal to the flow lines) is, by the Frobenius theorem, that the exterior 2-form

$$d\nu = \frac{1}{2}\,(u_{i;j} - u_{j;i})\, dx^j \wedge dx^i$$

be zero when restricted to each $u^\perp$. If we choose coordinates that are orthonormal at a given event and such that $\partial/\partial t$ is tangent to the flow at the event, then $d\nu$ restricted to $u^\perp$ (i.e., to $dx^0 = 0$) is

$$dv\big|_{u^\perp} = \frac{1}{2}(u_{\beta;\alpha} - u_{\alpha;\beta})\,dx^\alpha \wedge dx^\beta$$

$$= -\omega_{\alpha\beta}\,dx^\alpha \wedge dx^\beta = -2\omega_{\alpha<\beta}\,dx^\alpha \wedge dx^\beta.$$

Thus integrability of the distribution $u^\perp$ is equivalent to $\omega = 0$, that is, to vorticity zero.

It is a classical fact (which we shall employ only in the following example) that if α is any nonvanishing 1-form in an M^n, then the distribution of $(n-1)$-planes $\alpha = 0$ (i.e., the set of vectors X such that $\alpha(X) = 0$) is integrable if and only if $\alpha \wedge d\alpha = 0$. For example, consider the velocity field $\mathbf{v} = -y\,\partial/\partial x + x\,\partial/\partial y$ of a one-parameter family of rotations about the z-axis of R^3. The orthogonal 2-planes $-y\,dx + x\,dy = 0$ are integrable (with integral surfaces $y = kx$). On the other hand, the induced velocity field on $R \times R^3$, $u = -y\,\partial/\partial x + x\,\partial/\partial y + \partial/\partial t$ has an orthogonal 3-plane distribution $-y\,dx + x\,dy + dt = 0$ that is not integrable. In the case of our world-line flow in M^4, the unique vector field W such that $i_W\omega^4 = \nu \wedge d\nu$ is called the *vorticity vector*. It is not difficult to see that W is always orthogonal to u, i.e., W is a spatial vector.

Finally, we consider the expansion of the flow. Since u is the unit tangent,

$$\omega_\perp^3 = i_u\omega^4$$

is the 3-volume form for each $u^\perp$ (see p. 61ff). The flow is generated by $\tilde{u} = \varphi u$; for Lie derivative (p. 59) we obtain

$$\mathcal{L}_{\tilde{u}}\omega_\perp^3 = \mathcal{L}_{\tilde{u}} i_u\omega^4 = i_{\tilde{u}}\,d i_u\omega^4 = i_{\tilde{u}}\mathcal{L}_u\omega^4$$

$$= (\operatorname{div} u)\, i_{\tilde{u}}\omega^4,$$

or, since $\tilde{u} = u$ on $\mathscr{C}_0$,

$$\mathcal{L}_{\tilde{u}}\omega_\perp^3 = (\operatorname{div} u)\,\omega_\perp^3. \tag{7-14}$$

If the flow lines are considered the world lines of a fluid flow, we see that $\operatorname{div} u$ measures the logarithmic rate of volumetric expansion of the fluid in the ordinary three-dimensional sense. We have used the definition of divergence given by Equation 6-7. An equivalent definition for (pseudo)-Riemannian manifolds is (see Eqn. 10-7)

$$\operatorname{div} u = \text{trace of the linear transformation} \qquad Y \to \nabla_Y u.$$

If $\{e_0, e_1, e_2, e_3\}$ is an orthonormal frame, we then have

$$\begin{aligned} \operatorname{div} u &= \sum_{i=0}^{3} \varepsilon_i \langle \nabla_{e_i} u, e_i \rangle \\ &= -\langle \nabla_{e_0} u, e_0 \rangle + \sum_{\alpha=1}^{3} \langle \nabla_{e_\alpha} u, e_\alpha \rangle. \end{aligned} \tag{7-15}$$

In our case, with $u = e_0$,

$$\operatorname{div} u = \sum_{\alpha=1}^{3} \langle \nabla_{e_\alpha} u, e_\alpha \rangle = \sum_{\alpha=1}^{3} \theta_{\alpha\alpha} = \theta, \tag{7-16}$$

where θ is the trace of the form Θ (θ is also called, from Equation 7-14, the *expansion* of the fluid).

Shear and the Stress Tensor for a Viscous Fluid

Consider the world lines in M^4 corresponding to a fluid flow. Using again a Fermi-translated orthonormal basis along a world line $\mathscr{C}_0$, we have a relative velocity field in $u_0^\perp$ given by Equation 7-5:

$$v = \nabla_X u = u^i_{;j} X^j e_i = u^\alpha_{;\beta} X^\beta e_\alpha.$$

Thus, this velocity field is, at $t = 0$, the vector field on $u_0^\perp$ given by

$$\begin{cases} \mathbf{v}(\mathbf{x}) = \sum_{\alpha=1}^{3} v_\alpha(\mathbf{x}) \mathbf{e}_\alpha \\ v_\alpha(\mathbf{x}) = \sum_{\beta=1}^{3} u_{\alpha;\beta}(0) x_\beta \end{cases} \tag{7-17}$$

where $[u_{\alpha;\beta}(0)]$ is the matrix $u_{\alpha;\beta}$ at the particular event $\mathscr{C}(0)$. The deformations of the fluid are described by the matrix $u_{\alpha;\beta}$. Now

$$u_{\alpha;\beta} = \theta_{\alpha\beta} + \omega_{\alpha\beta}$$

and $(\omega_{\alpha\beta})$ generates rotations, i.e., rigid motions of the fluid. It is usual to assume that the three-dimensional stress tensor for the fluid $(\tilde{S}_{\alpha\beta})$ is independent of $\omega_{\alpha\beta}$; isometries set up no stresses in a fluid.

Any symmetric matrix, such as Θ, can be further decomposed in an invariant fashion into an isotropic part plus a trace-free part:

$$\Theta = \frac{1}{3}\,\mathrm{tr}\,\Theta\,\mathsf{I} + \sigma = \frac{\theta}{3}\,\mathsf{I} + \sigma.$$

The trace-free part σ of the rate-of-strain matrix Θ,

$$\sigma_{\alpha\beta} = \theta_{\alpha\beta} - \frac{1}{3}\,\theta\delta_{\alpha\beta}, \tag{7-18}$$

is called the *rate of shear* form.

The simplest three-dimensional stress tensor $\tilde{\mathsf{S}}$ for a viscous fluid is of the form (recall that $\tilde{\mathsf{S}}$ *is* symmetric)

$$\tilde{S}_{\alpha\beta} = 2\eta\sigma_{\alpha\beta} + b\theta\delta_{\alpha\beta} - p\delta_{\alpha\beta}.$$

Here p is an isotropic pressure, η is a coefficient of shear viscosity, and b is another viscosity parameter. The simplest viscous fluids show no viscous stresses under uniform expansions. For uniform expansion $\theta_{\alpha\beta} = \dot{P}_{\alpha\beta}(0) = \lambda\delta_{\alpha\beta}$ (see Eqn. 7-13) and so $\theta = 3\lambda$ and $\sigma_{\alpha\beta} = 0$. Assuming then that there are no viscous stresses in uniform expansion, we have $b\theta\delta_{\alpha\beta} = 0$, and consequently

$$\tilde{S}_{\alpha\beta} = 2\eta\sigma_{\alpha\beta} - p\delta_{\alpha\beta} \tag{7-19}$$

is the appropriate stress tensor. Note, from Equation 7-17 we have $u_{\alpha;\beta} = \partial v_\alpha/\partial x^\beta$ (and consequently the integrability of $u^\perp$ is equivalent to the vanishing of the curl of the relative velocity vector $\mathbf{v}$). We then have the classical form

$$\tilde{S}_{\alpha\beta} = \eta\left[\left(\frac{\partial v_\alpha}{\partial x_\beta} + \frac{\partial v_\beta}{\partial x_\alpha}\right) - \frac{2}{3}\mathrm{div}\,\mathbf{v}\,\delta_{\alpha\beta}\right] - p\delta_{\alpha\beta}.$$

We now construct the associated four-dimensional stress-energy-momentum tensor for the fluid. First, let $\tilde{\mathsf{S}}: u_0^\perp \times u_0^\perp \to R$ be the bilinear form associated with $(\tilde{S}_{\alpha\beta})$. Let $\Pi: M^4{}_{\mathscr{C}_0(0)} \to M^4{}_{\mathscr{C}_0(0)}$ be orthogonal projection onto $u_0^\perp$. Thus $\Pi Y = Y + \langle Y, u\rangle u$, and in terms of any coordinates Π has matrix

$$\Pi^i_j = \delta^i_j + u^i u_j. \tag{7-20}$$

The four-dimensional covariant stress tensor S is then defined by

$$\mathsf{S}(X, Y) = \tilde{\mathsf{S}}(\Pi X, \Pi Y).$$

The covariant mass tensor M is defined by

$$\mathsf{M} = \rho_0 \nu \otimes \nu$$

or

$$M_{ij} = \rho_0 u_i u_j,$$

where ρ_0 is the mass density as measured by a co-moving observer, i.e., an observer whose 4-velocity at the given event is u. Then the covariant stress-energy-momentum tensor for the fluid is defined by

$$\begin{aligned} \mathsf{T}(X, Y) &= \mathsf{M}(X, Y) - \mathsf{S}(X, Y) \\ &= \rho_0 \langle u, X \rangle \langle u, Y \rangle - \tilde{\mathsf{S}}(\Pi X, \Pi Y). \end{aligned}$$

For a viscous fluid $\tilde{\mathsf{S}}(\Pi X, \Pi Y) = -p \langle \Pi X, \Pi Y \rangle + 2\eta\sigma(\Pi X, \Pi Y)$. In the case of a perfect fluid, $\eta = 0$,

$$\begin{aligned} \mathsf{T}(X, Y) &= \rho_0 \langle u, X \rangle \langle u, Y \rangle + p \langle X + \langle u, X \rangle u, Y + \langle u, Y \rangle u \rangle \\ &= (\rho_0 + p) \langle u, X \rangle \langle u, Y \rangle + p \langle X, Y \rangle. \end{aligned}$$

In terms of any coordinates, $\mathsf{T}(X, Y) = T_{ij} X^i Y^j$ and $T_{ij} = (\rho_0 + p) u_i u_j + p g_{ij}$, which is the form of the stress-energy-momentum tensor used in the derivation of the Einstein equations (p. 32).

The stress-energy-momentum tensor for general relativity will be assumed to be *symmetric:* $T_{ij} = T_{ji}$. The case of the electromagnetic field will be discussed in Chapter 10.

Divergence of the Einstein Tensor: Gravitational "Force"

The classical equations of motion in their four-dimensional form involve the vector with components $\partial T^{ij}/\partial x^j$ (see Eqn. 6-22). We expect that the relativistic equations of motion will involve the vector with components $T^{ij}{}_{;j}$, i.e., the divergence of the stress-energy-momentum tensor. While the classical equations (Eqn. 6-19) were obtained via the integral laws of Chapter 6 (Eqn. 6-18) there is no hope of proceeding in this fashion in general relativity *since the integral of a vector quantity makes no invariant* (i.e., coordinate-independent) *sense in a curved manifold;* one simply cannot add vectors based at different points. Nevertheless, we shall now see that the Einstein equations $G^{ij} = 8\pi\kappa T^{ij}$ do yield equations that can be interpreted as equations of motion. Since $G^j{}_i = R^j{}_i - ½\, \delta^j_i R$, we have, on taking the divergence of both sides, $G^j{}_{i;j} = R^j{}_{i;j} - \frac{1}{2}\, \delta^j_i \frac{\partial R}{\partial x^j}$ or

$$G^j{}_{i;j} = R^j{}_{i;j} - \frac{1}{2}\frac{\partial R}{\partial x^i}. \tag{7-21}$$

On the other hand, from the classical Bianchi identities

$$R^j{}_{ikl;m} + R^j{}_{imk;l} + R^j{}_{ilm;k} = 0,$$

put $k = j$ and sum over j:

$$R_{il;m} - R_{im;l} + R^j{}_{ilm;j} = 0.$$

Multiply by g^{ri} and sum over i:

$$R^r{}_{l;m} - R^r{}_{m;l} + R^{jr}{}_{lm;j} = 0.$$

Put $m = r$ and sum over r, using $R^r{}_{r;l} = R_{;l} = \partial R/\partial x^l$:

$$R^r{}_{l;r} - \frac{\partial R}{\partial x^l} + R^j{}_{l;j} = 0.$$

This gives us the classical fact

$$\frac{\partial R}{\partial x^l} = 2R^j{}_{l;j}. \tag{7-22}$$

Combining this with Equation 7-21, we conclude that the Einstein tensor has divergence zero:

$$G^j{}_{i;j} = 0,$$

and consequently

$$T^j{}_{i;j} = 0 \qquad (i = 0, 1, 2, 3) \tag{7-23}$$

is an automatic consequence of the Einstein equations! Every tensor proposed as the full stress-energy-momentum tensor in general relativity must have this property of being divergence free.

Consider, for example, a fluid subject only to gravitational forces. Comparing Equation 7-23 with the corresponding classical equations (Eqn. 6-22), we see that Equation 7-23 corresponds to the case of zero external force; *gravitation is not to be considered as an external force in the space-time manifold!* A point particle undergoing free fall in a pure gravitational field is not subject to any "force" of gravity. This will be further clarified in the next section.

The Equations of Motion

If A is any contravariant tensor of order $k > 0$, its divergence $\mathsf{Div\ A}$ is the contravariant tensor of order $k - 1$ whose components are given by

$$(\mathsf{Div\ A})^{i_1 \cdots i_{k-1}} = A^{i_1 \cdots i_{k-1} j}{}_{;j}.$$

Consider now the equations $T^{ij}{}_{;j} = 0$ for a stress-energy-momentum tensor of the form

$$T^{ij} = \rho_0 u^i u^j - S^{ij},$$

where u is the unit tangent to a flow on M^4. Then

$$0 = T^{ij}{}_{;j} = (\rho_0 u^j) u^i{}_{;j} + (\rho_0 u^j)_{;j} u^i - S^{ij}{}_{;j}$$

or

$$0 = \rho_0 \nabla_u u + \mathsf{div}(\rho_0 u) u - \mathsf{Div\ S}.$$

Since the geodesic curvature $\nabla_u u$ of the world line is orthogonal to the unit tangent u we may write this as "equations of motion":

$$\begin{cases} \mathsf{div}(\rho_0 u) = -\langle \mathsf{Div\ S}, u \rangle \\ \rho_0 \nabla_u u = \Pi\ \mathsf{Div\ S} \end{cases} \tag{7-24}$$

where as usual Π is projection onto $u^\perp$.

In the case of a perfect fluid, from $\tilde{\mathsf{S}} = -p\mathsf{I}$ and Equation 7-20 we see

$$S^{ij} = -p(g^{ij} + u^i u^j)$$

and so

$$(\text{Div S})^i = S^{ij}{}_{;j} = -p_{;j}(g^{ij} + u^i u^j) - p(u^i{}_{;j} u^j + u^i u^j{}_{;j})$$

or

$$\text{Div S} = -\Pi \text{ grad } p - p\nabla_u u - p(\text{div } u)\cdot u.$$

Equation 7-24 then becomes

$$\boxed{\begin{aligned} \text{div } \rho_0 u &= -p \text{ div } u \\ (\rho_0 + p)\nabla_u u &= -\Pi \text{ grad } p = -(\text{grad } p)^{\perp} \end{aligned}}\qquad \begin{aligned}(7\text{-}25)\\(7\text{-}26)\end{aligned}$$

where $(\text{grad } p)^{\perp} = \Pi \text{ grad } p$ is the component of grad p normal to u.

Equation 7-25 replaces the classical equation of conservation of mass (Eqn. 6-16), for if we again let $\omega_{\perp}^3 = i_u \omega^4$ be the 3-volume form in each $u^{\perp}$, then as in Equation 7-14

$$\mathcal{L}_u \omega_{\perp}^3 = \text{div } u \;\; \omega_{\perp}^3,$$

and for rate of change of rest mass (i.e., mass as measured by a co-moving observer)

$$\begin{aligned} \mathcal{L}_u(\rho_0 \omega_{\perp}^3) &= u(\rho_0)\omega_{\perp}^3 + \rho_0 \mathcal{L}_u \omega_{\perp}^3 \\ &= u(\rho_0)\omega_{\perp}^3 + \rho_0(\text{div } u)\omega_{\perp}^3 \end{aligned}$$

so that

$$\mathcal{L}_u(\rho_0 \omega_{\perp}^3) = \text{div}(\rho_0 u)\omega_{\perp}^3, \qquad (7\text{-}27)$$

and Equation 7-25 then says

$$\mathcal{L}_u(\rho_0 \omega_{\perp}^3) = -p\mathcal{L}_u(\omega_{\perp}^3). \qquad (7\text{-}28)$$

Classically, $-p\mathcal{L}_u \omega_{\perp}^3$ represents the rate at which pressure does work in opposing the expansion of the fluid. Equation 7-25 then says that mass is not conserved; work done by pressure in opposing expansion contributes to the energy content, hence mass.

Consider next Equation 7-26. The geodesic curvature vector $\nabla_u u$ generalizes the acceleration vector $du/d\tau$ of special relativity. Many authors call $\nabla_u u$ the *acceleration vector* along the world line. Equation

7-26 is then a four-dimensional analog of the Euler equation $\rho\, d\mathbf{v}/dt = \mathbf{b} - \operatorname{grad} p$ (see p. 68). We have remarked earlier that gravitation is not to be considered as an external force, that is, **b** is to be put equal to zero. We can see how general relativity accomplishes this, as follows. In classical gravitation $\mathbf{b} = \operatorname{grad} U$ is the gradient of the gravitational potential. The gravitational potential U has been generalized and replaced by the tensor potentials (g_{ij}), which are incorporated into the geometric structure of space-time itself. The left side of Equation 7-26 involves $\nabla_u u$, which involves the Christoffel symbols (see Chapter 4), which in turn may be considered as generalized gradients of the potentials (g_{ij}). Thus the terms that classically would correspond to $\mathbf{b} = \operatorname{grad} U$ have been embedded into the left side of Equation 7-26!

The contribution of p in the left side of Equation 7-26 is ordinarily very small. If we had used c instead of 1 for the velocity of light, $\rho_0 + p$ would have been replaced by $\rho_0 + p/c^2$.

An especially simple but highly idealized case is that of an *incoherent fluid,* or a *dust,* i.e., the case in which the pressure vanishes ($p = 0$). In this case Equations 7-25 and 7-26 say that *mass is conserved* and *the world lines of the dust are geodesics!*

A more realistic fluid body, e.g., an idealized model of a planet moving in the solar system, will not have its molecules moving along geodesic world lines because of the pressure gradient. If there were a "central molecule" where the pressure was always a spatial maximum, then its world line would indeed be a geodesic, but such a "center" has never been proven to exist. (For related discussions see Synge, 1960, Chapter 4, §7 and Chapter 6, §6.)

It is important to realize that when discussing the curvature of a world line of a point in a body, the body itself influences the geometry of space time via Einstein's equations. We can envision the body becoming increasingly smaller and its influence on the metric becoming relatively insignificant. A *test particle* is a mathematical abstraction of this situation where the particle is assumed not to influence at all the "backround metric," i.e., the metric due to all other matter and energy in the universe. The case of an incoherent dust encourages us to postulate the following geodesic hypothesis.

> A test particle subject to no nongravitational "forces" has a geodesic world line.

A *light ray* is the world line of a photon; we shall also assume that this is a geodesic, in this case a *null-geodesic*. There are indications that the geodesic hypothesis is in fact a mathematical consequence of Einstein's equations, but this is apparently a very delicate matter. (For discussion see, for example, Bergmann, 1942.)

Geodesics and Constants of Motion

Recall now the basic facts about geodesics (for details I recommend Milnor, 1963, especially §12).

Let $\mathscr{C}_\varepsilon$ be a one-parameter family of curves, each parameterized by a given parameter $\lambda \in [\lambda_1, \lambda_2]$. Each $\mathscr{C}_\varepsilon$ is a *variation* of the base curve $\mathscr{C}_0$. The tangent vector field along $\mathscr{C}_\varepsilon$ will be written as $T(\varepsilon, \lambda) = \partial x/\partial\lambda$. It has become traditional in mathematics to define the *energy,* or *action,* of the curve $\mathscr{C}_\varepsilon$ by

$$A(\varepsilon) = \frac{1}{2} \int_{\lambda_1}^{\lambda_2} \langle T(\varepsilon, \lambda), T(\varepsilon, \lambda)\rangle \, d\lambda$$

but I should emphasize that no "physical" energy is associated with this integral as used here. Using the basic relation between the tangent and variation vectors (Milnor, 1963),

$$\frac{\nabla T}{\partial \varepsilon} = \frac{\nabla X}{\partial \lambda},$$

which is classically written as $\delta\, dx = d\delta x$, we compute the *first variation of energy,* as follows:

$$\begin{aligned} 2A'(\varepsilon) &= \frac{d}{d\varepsilon} \int_{\lambda_1}^{\lambda_2} \langle T, T\rangle d\lambda = \int_{\lambda_1}^{\lambda_2} \frac{\partial}{\partial \varepsilon} \langle T, T\rangle d\lambda \\ &= 2 \int_{\lambda_1}^{\lambda_2} \left\langle \frac{\nabla T}{\partial \varepsilon}, T \right\rangle d\lambda = 2 \int_{\lambda_1}^{\lambda_2} \left\langle \frac{\nabla X}{\partial \lambda}, T \right\rangle d\lambda \\ &= 2 \int_{\lambda_1}^{\lambda_2} \frac{\partial}{\partial \lambda} \langle X, T\rangle d\lambda - 2 \int_{\lambda_1}^{\lambda_2} \left\langle X, \frac{\nabla T}{\partial \lambda} \right\rangle d\lambda \end{aligned}$$

$$A'(\varepsilon) = \langle X, T\rangle \Big|_{\lambda_1}^{\lambda_2} - \int_{\lambda_1}^{\lambda_2} \left\langle X, \frac{\nabla T}{\partial \lambda} \right\rangle d\lambda. \tag{7-29}$$

Then $A'(0)$ is the first variation. $\mathscr{C}_0 = \mathscr{C}$ is said to be a geodesic if $A'(0) = 0$ for all variations whose variation vector $X(0, \lambda)$ vanishes at $\lambda = \lambda_1$ and $\lambda = \lambda_2$. From $A'(0)$ we get by the usual argument that $\mathscr{C}$ must satisfy the geodesic equation

$$\nabla_T T \equiv \frac{\nabla T}{\partial \lambda} = 0$$

or

$$\frac{d^2x^i}{d\lambda^2} + \Gamma^i{}_{jk} \frac{dx^j}{d\lambda} \frac{dx^k}{d\lambda} = 0.$$

Along such a geodesic we have

$$\frac{d}{d\lambda} \langle T, T \rangle = 2 \langle \nabla_T T, T \rangle = 0,$$

and so T must have constant length along $\mathscr{C}$. If $\mathscr{C}$ is time-like (resp. space-like) the parameter must be proportional to proper time (resp. arc length). If $\mathscr{C}$ is light-like, i.e., $\langle T, T \rangle = 0$, the parameter λ is still determined up to a change of the form $\lambda' = a\lambda + b$, as follows easily from the geodesic equations. Any parameter λ such that $\nabla_T T = 0$ will be called a *distinguished parameter* or *affine parameter* for the geodesic.

The geodesic equations can seldom be solved exactly. Frequently it is important to have information about the geodesic corresponding to the "constants of motion" of classical dynamics. For this purpose consider the following. Let $\varphi_\varepsilon : M \to M$ be a one-parameter group of *isometries* of M. The velocity vector field

$$X(p) = \frac{\partial}{\partial \varepsilon} (\varphi_\varepsilon p) \bigg|_{\varepsilon=0}$$

is then a Killing vector field. If $\mathscr{C}$ is a given curve in M, we can construct a variation of $\mathscr{C}$ by $\mathscr{C}_\varepsilon = \varphi_\varepsilon(\mathscr{C})$, with $\mathscr{C}_0 = \mathscr{C}$. Let $\mathscr{C}$ be a geodesic parameterized by distinguished parameter λ. Since each φ_ε is an isometry, $\mathscr{C}_\varepsilon$ is again a geodesic parameterized by distinguished parameter λ and $A(\varepsilon) = A(0) =$ constant. From the first variation formula (Eqn. 7-29) and $\nabla_T T = 0$ we conclude (since λ_1 and λ_2 are arbitrary parameter values for $\mathscr{C}$) the classical constant of motion lemma.

Constant of motion lemma If X is a Killing vector field and $T = dx/d\lambda$ is the tangent vector to a geodesic $\mathscr{C}$ of M with distinguished parameter λ, then $\langle X, T \rangle$ is constant along $\mathscr{C}$.

As an application, consider the motion of a planet about the sun. We shall neglect all other matter in the universe. We shall also assume that the planet is so small that we may consider it a test particle; by the geodesic hypothesis the world line of the planet is a geodesic in the back-round metric, which is merely the Schwarzschild metric with the sun at origin:

$$ds^2 = -\left(1 - \frac{2m}{r}\right)dt^2 + \frac{dr^2}{1 - \dfrac{2m}{r}} + r^2(d\theta^2 + \sin^2\theta\, d\varphi^2),$$

where m is the sun's mass.

Let $\lambda = \tau$ be proper time, i.e., the distinguished parameter along this time-like geodesic. For tangent 4-vector we have

$$T = \frac{dt}{d\tau}\frac{\partial}{\partial t} + \frac{dr}{d\tau}\frac{\partial}{\partial r} + \frac{d\theta}{d\tau}\frac{\partial}{\partial \theta} + \frac{d\varphi}{d\tau}\frac{\partial}{\partial \varphi}, \qquad \langle T, T\rangle = -1.$$

Since the Schwarzschild metric is independent of t and φ, it follows that $\partial/\partial t$ and $\partial/\partial\varphi$ are Killing vector fields. We then have the following constants of motion:

$$\begin{cases} \left\langle T, \dfrac{\partial}{\partial t}\right\rangle = \dfrac{dt}{d\tau}\left\langle \dfrac{\partial}{\partial t}, \dfrac{\partial}{\partial t}\right\rangle = \dfrac{dt}{d\tau} g_{00} = -\dfrac{dt}{d\tau}\left(1 - \dfrac{2m}{r}\right) = \text{const.} = -E \\ \left\langle T, \dfrac{\partial}{\partial \varphi}\right\rangle = \dfrac{d\varphi}{d\tau}\left\langle \dfrac{\partial}{\partial \varphi}, \dfrac{\partial}{\partial \varphi}\right\rangle = \dfrac{d\varphi}{d\tau} g_{\varphi\varphi} = \dfrac{d\varphi}{d\tau} r^2\sin^2\theta = \text{const.} = h. \end{cases} \tag{7-30}$$

Suppose, for example, the planet has an initial spatial velocity vector tangent to the spatial surface $\theta = \pi/2$. By symmetry the planet will remain on this surface. The second constant of motion gives $r^2\, d\varphi/d\tau =$ constant $= h$, which replaces the classical angular momentum "constant" $r^2 d\varphi/dt$. This classical quantity is not in general a constant in general relativity, since

$$r^2\frac{d\varphi}{dt} = r^2\frac{d\varphi}{d\tau}\frac{d\tau}{dt} = \frac{h}{E}\left(1 - \frac{2m}{r}\right),$$

which is not a constant except in circular motion.

Next, consider radial motion of such a planet. It is easy to see that such motions exist from the "integral" $r^2 d\varphi/d\tau = h$, for if the motion is initially radial, i.e., $d\varphi/d\tau = 0$ at initial time t, then φ must be constant. Since $d\theta = 0 = d\varphi$, we have

$$ds^2 = g_{00}(r)\, dt^2 + g_{rr}(r)\, dr^2,$$

and since ds^2 is negative for time-like curves, we have

$$-\left(\frac{d\tau}{dt}\right)^2 = g_{00}(r) + g_{rr}(r)\left(\frac{dr}{dt}\right)^2,$$

and so

$$\left(\frac{dr}{dt}\right)^2 = \frac{-g_{00}}{g_{rr}}\left(\frac{g_{00}}{E^2} + 1\right) = \left(1 - \frac{2m}{r}\right)^2\left[1 - \frac{\left(1 - \frac{2m}{r}\right)}{E^2}\right].$$

Consider a particle released with initial coordinate speed $dr/dt = 0$ at $r = R > 2m$. Then $E^2 = (1 - 2m/R)$ and a fixed observer would note that *the particle would take an infinite coordinate time t to reach the Schwarzschild singularity* (an easy consequence of $dr/dt \to 0$ as $r \to 2m$). On the other hand, an observer falling with the particle and using his proper time τ would observe

$$\left(\frac{dr}{d\tau}\right)^2 = \left(\frac{dr}{dt}\right)^2\left(\frac{dt}{d\tau}\right)^2 = E^2 - \left(1 - \frac{2m}{r}\right) \to 1 - \frac{2m}{R}$$

as $r \to 2m$; it takes but a finite amount of *proper* time to "cross the singularity." (For more details, see, for example, Misner, Thorne, and Wheeler, 1973, Chapter 25.)

The general relativistic treatment of the orbit of a planet about the sun leads to Einstein's famous explanation of the precession of the classical elliptical orbit. The analysis is based on the constants of motion (Eqn. 7-30) and is treated in most books on relativity (see, for example, Adler, Bazin, and Schiffer, 1975).

The three famous "tests" of relativity are the red shift formula (Eqn. 2-4), the precession of Mercury's orbit mentioned above, and the bending of a light ray near the surface of the sun, to be discussed in our next chapter.

Tidal Forces

Earlier in this chapter we defined the relative position vector $\mathbf{x} = X$ and the relative velocity vector $\mathbf{v} = \nabla_X u = \mathsf{F}_u X$ for a congruence of world lines based on a world line $\mathscr{C}_0$. We can define a *relative acceleration* vector by $\mathbf{a} = \mathsf{F}_u \mathsf{F}_u X$. When $\mathscr{C}_0$ is a geodesic, Fermi differentiation coincides with covariant differentiation, and we then have for acceleration

$\mathbf{a} = \nabla_u \nabla_u X$. Consider a cloud of dust particles; their world lines are geodesics, and we shall let $\mathscr{C}_0$ be a base dust particle world line. For relative acceleration we have (essentially from Equations 4-1, 7-4, and 7-5)

$$\mathbf{a} = \nabla_u \nabla_u X = \nabla_u \nabla_X u = \nabla_X(\nabla_u u) + \mathsf{R}(u, X)u$$
$$= -\mathsf{R}(X, u)u.$$

The transformation $X \to -\mathsf{R}(X, u)u$ is easily seen, from the symmetry properties of the Riemann tensor, to be a self-adjoint linear transformation of $u^\perp$ into itself. The acceleration $\mathbf{a}(X)$ is called the "tidal" acceleration at X and is classically explained in terms of inhomogeneities of the gravitational field. Consider, for example, an idealized planet consisting of a dust drop revolving slowly about the sun. By Eqn. 7-26 the center of the drop describes a geodesic world line in this space-time that (neglecting the influence of the drop) is assumed to be the Schwarzschild solution exterior to the sun. The velocity vector u of the world lines is very nearly $e_t = (1 - 2m/r)^{-1/2}\ \partial/\partial t$, since its spatial component is assumed small. From geometric symmetry it is clear that $-\mathsf{R}(e_r, e_t)e_t$ must lie along the radial direction. Thus for the radial component of $\mathbf{a}$ at position $X = \xi e_r$ we have (using Eqn. 5-19)

$$\langle \mathbf{a}(\xi e_r), e_r \rangle = -\langle \mathsf{R}(e_r, e_t)e_t, e_r \rangle \xi$$
$$= \xi \mathsf{K}(e_r \wedge e_t) = \frac{2m}{r^3}\, \xi.$$

Also, since $X \to -\mathsf{R}(X, u)u$ is self-adjoint, it follows that $\mathbf{a}(e_\theta)$ is directed along e_θ, and

$$\langle \mathbf{a}(\xi e_\theta), e_\theta \rangle = \xi \mathsf{K}(e_\theta \wedge e_t) = -\frac{m}{r^3}\, \xi.$$

Thus, if the drop were composed of *dust* particles, the particle at distance ξ along the e_r direction from the center would accelerate away from the center with acceleration $(2m/r^3)\xi$, while the particles in the "plane" through the center orthogonal to e_r would suffer an acceleration $(m/r^3)\xi$ toward the central particle. In a *fluid* drop, however, these accelerations are resisted by the (nongravitational) molecular forces, resulting in tidal bulges elongating the drop in the e_r direction.

8

Light Rays and Fermat's Principle

Fermat's Principle of Stationary Time

A coordinate system t, x^1, x^2, x^3 decomposes space-time into "spatial slices," t = constant. According to the geodesic hypothesis of the previous chapter, a light ray traces out a geodesic world line in space-time M^4. The spatial trace of the light ray need not, however, be a geodesic in the spatial metric, and we shall investigate the spatial curvature of the ray in this chapter.

Recall, as in Equation 2-3, that if the space-time metric is of the form $ds^2 = g_{00}\,dt^2 + dl^2$, where dl^2 contains no dt terms, then since the path of a light ray satisfies $ds^2 = 0$, we have $dl/dt = \sqrt{-g_{00}}$, i.e., the light ray has a spatial *coordinate* speed of $\sqrt{-g_{00}}$. In particular, in a static universe, this speed decreases when the gravitational potential $U = 1 - \sqrt{-g_{00}}$ increases. This gives rise, via Huygen's principle, to a formula for the bending of light rays near a massive body. This was first done by Einstein in 1911, but his analysis was incorrect since he was still working in the context of a flat Minkowski space. We shall proceed along different lines, via Fermat's principle of "least" time, and then return to Einstein's 1911 result.

Consider a static universe $ds^2 = g_{00}\, dt^2 + dl^2$, and let $\tilde{P}$ and $\tilde{Q}$ be nearby points in the spatial section $V^3 = V^3_0$ (Figure 8-1). Consider a one-parameter family of curves $\tilde{\mathscr{C}}_\epsilon$ joining $\tilde{P}$ to $\tilde{Q}$, and suppose that each of these curves is traversed with the speed of light, $dl/dt = \sqrt{-g_{00}}$. We wish to study the time necessary to reach $\tilde{Q}$ along these curves $\tilde{\mathscr{C}}_\epsilon$. Geometrically then, each of the curves $\tilde{\mathscr{C}}_\epsilon$ has a unique lift to a light-like curve $\mathscr{C}_\epsilon$, all starting say at $P = (0, \tilde{P})$, but ending perhaps at different times t_ϵ over $\tilde{Q}$. Parameterize each of these curves $\mathscr{C}_\epsilon$ by a parameter λ, $0 \leq \lambda \leq 1$, to be specified shortly. For "energy" we have

$$2A(\varepsilon) = \int_0^1 \left\langle \frac{\partial x}{\partial \lambda}, \frac{\partial x}{\partial \lambda} \right\rangle d\lambda = \int_0^1 \langle T, T \rangle \, d\lambda = 0,$$

since each $\mathscr{C}_\epsilon$ is light-like. Thus, for first variation (Eqn. 7-29),

$$0 = A'(0) = \langle X, T \rangle \Big|_P^Q - \int_0^1 \langle X, \nabla_T T \rangle \, d\lambda.$$

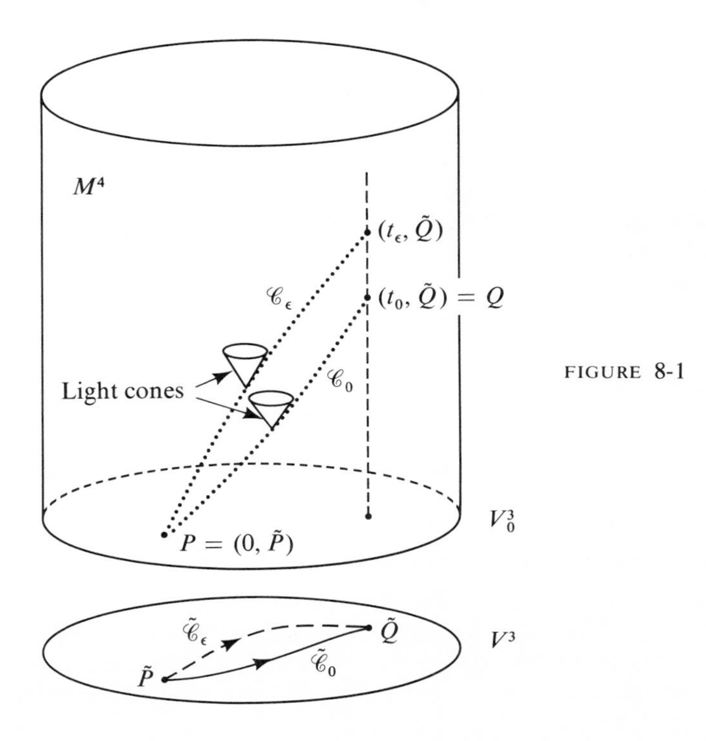

FIGURE 8-1

By construction of the lifted curves, $X = \partial x/\partial \varepsilon = 0$ at P and also X has no spatial component at Q; let us define (using classical notation) δt by $X = \delta t \cdot \partial/\partial t$ at Q. Then the first variation formula gives

$$\delta t = \left\langle \frac{\partial}{\partial t}, T \right\rangle^{-1} \int_0^1 \langle X, \nabla_T T \rangle \, d\lambda,$$

and since T is light-like, $\langle \partial/\partial t, T \rangle$ is not zero. Now let $\mathscr{C}_0$ be a light ray, i.e., $\tilde{\mathscr{C}}_0$ is the spatial path of a light ray from $\tilde{P}$ to $\tilde{Q}$. Choose now for λ a distinguished parameter along $\mathscr{C}_0$. Then $\nabla_T T = 0$ and we have proved "*Fermat's principle*" of stationary time, $\delta t = 0$, *i.e., the spatial path of a light ray gives an extremal for the time necessary to go from $\tilde{P}$ to $\tilde{Q}$ while traveling at the (local) speed of light* $\sqrt{-g_{00}}$.

We can express Fermat's principle in the form (using classical notation)

$$\delta \int dt = \delta \int \frac{dl}{\sqrt{-g_{00}}} = 0$$

for the spatial trace of the light ray. Thus the spatial path of a light ray is a geodesic in V^3, not in the spatial metric dl but rather in the "Fermat metric"

$$dl_F = \frac{dl}{\sqrt{-g_{00}}}.$$

Perhaps it is worth pointing out exactly what the distinguished parameter λ for a light-like geodesic $\mathscr{C}$ is in this static case. From the constant of motion lemma (p. 86),

$$\left\langle \frac{\partial}{\partial t}, T \right\rangle = \left\langle \frac{\partial}{\partial t}, \frac{dx}{d\lambda} \right\rangle = \text{constant, say } k.$$

Since $ds^2 = g_{00}\, dt^2 + dl^2$ has no $dt\, dx^\alpha$ terms,

$$k = \left\langle \frac{\partial}{\partial t}, \frac{dt}{d\lambda} \frac{\partial}{\partial t} \right\rangle = g_{00} \frac{dt}{d\lambda},$$

i.e., $d\lambda/dt = g_{00}/k$ along $\mathscr{C}$. Thus, along $\mathscr{C}$

$$\frac{dl}{d\lambda} = \frac{dl}{dt} \cdot \frac{k}{g_{00}} = \sqrt{-g_{00}} \cdot \frac{k}{g_{00}} = \frac{-k}{\sqrt{-g_{00}}}.$$

Thus the distinguished parameter λ is such that one moves along the spatial path $\tilde{\mathscr{C}}$ with λ speed $dl/d\lambda$ inversely proportional to the local speed of light $\sqrt{-g_{00}}$!

Geodesics in Conformally Related Metrics

The spatial path of a light ray is a geodesic in the Fermat metric dl_F; how is it "curved" in the induced spatial metric dl? These metrics are conformally related (in particular, angular measurements are the same in both metrics). The following purely mathematical lemma will give an important piece of information for our problem. First, recall the following notations. If f is a function and N is a vector, then $N(f) = \langle N, \operatorname{grad} f\rangle = df(N)$ denotes the derivative of f with respect to N. Second, if T is the unit vector field tangent to a curve then $\nabla_T T = k_g N$, where k_g is the *geodesic curvature* for the curve and N is the unit *principal normal* vector to the curve.

Lemma Let dl and $dl_F = f\,dl$ be conformally related Riemannian metrics for a manifold. Let $\mathscr{C}$ be a dl_F-geodesic. Then the geodesic curvature of $\mathscr{C}$ in the dl-metric satisfies $k_g = N(\log f)$, where N is the dl-unit principal normal vector to $\mathscr{C}$, when $k_g \neq 0$.

PROOF We compute the first variation of arc length of $\mathscr{C}$. Consider then a variation $x = x(\varepsilon, \lambda)$ of $\mathscr{C} = \mathscr{C}_0$, where λ reduces to arc length dl on the base curve $\mathscr{C}$; $x = x(0, \lambda)$, and where the variation vanishes at the end points $\lambda = 0$ and $\lambda = b$. $T = \partial x/\partial\lambda$ is the tangent vector field to the varied curves, $X = \partial x/\partial\varepsilon$ is the variation vector field, and we shall let $\langle,\rangle$ and ∇ refer to the dl metric. Then for dl_F-arc length of $\mathscr{C}_\varepsilon$ we have

$$L_F(\varepsilon) = \int f(\varepsilon, \lambda)\,dl = \int_0^b f(\varepsilon, \lambda)\langle T, T\rangle^{1/2}\,d\lambda$$

and

$$\begin{aligned} L_F'(\varepsilon) &= \int_0^b \frac{\partial}{\partial\varepsilon}\,[f\cdot\langle T, T\rangle^{1/2}]\,d\lambda \\ &= \int_0^b \left[X(f)\langle T, T\rangle^{1/2} + f\cdot\frac{\langle \nabla T/\partial\varepsilon, T\rangle}{\langle T, T\rangle^{1/2}}\right]d\lambda. \end{aligned}$$

Using again $\nabla T/\partial\varepsilon = \nabla X/\partial\lambda$ and the fact that $\langle T, T\rangle = 1$ when $\varepsilon = 0$, we get, since $L_F(\varepsilon) = \int dl_F$ and $\mathscr{C}_0$ is a dl_F-geodesic,

$$0 = L_F'(0) = \int_0^b \left[X(f) + f \cdot \left\langle \frac{\nabla X}{\partial\lambda}, T\right\rangle\right] d\lambda$$

$$= \int_0^b \left[X(f) + f \cdot \left\{\frac{\partial}{\partial\lambda}\langle X, T\rangle - \langle X, \nabla_T T\rangle\right\}\right] d\lambda.$$

Choose for variation vector $X(\lambda) = \varphi(\lambda)\nabla_T T$, where φ is any smooth function vanishing at the endpoints $\lambda = 0$ and $\lambda = b$. Since $\nabla_T T$ is orthogonal to T along $\mathscr{C}$,

$$0 = \int_0^b \varphi(\lambda)[k_g N(f) - f \cdot k_g^2]\, d\lambda,$$

and since this is true for all such φ, we conclude $k_g N(f) = f k_g^2$, as desired.

Before proceeding further, I wish to make two purely mathematical remarks. First, exactly the same procedure will work for *minimal varieties* of higher dimensions, i.e., if $V^p \subset M^n$ is a dl_F-minimal submanifold, then its dl-mean curvature vector HN satisfies $H = N(\log f)$.

Second, consider the application of our lemma to a famous case. The *Poincaré metric* in the upper half plane is given by

$$dl_F = \frac{dl}{y} = \frac{\sqrt{dx^2 + dy^2}}{y}, \qquad y > 0,$$

i.e., $f(x, y) = 1/y$. Let $\mathscr{C}$ be a geodesic of the Poincaré metric. Then the Euclidean curvature of $\mathscr{C}$ is given by

$$k_g = N\left(\log \frac{1}{y}\right) = -N(\log y)$$

$$= -\left\langle N, \frac{\partial}{\partial x}\right\rangle \frac{\partial}{\partial x}(\log y) - \left\langle N, \frac{\partial}{\partial y}\right\rangle \frac{\partial}{\partial y}(\log y)$$

$$= -\frac{1}{y}\left\langle N, \frac{\partial}{\partial y}\right\rangle.$$

A rotation through 90° shows $\langle N, \partial/\partial y\rangle = \langle T, \partial/\partial x\rangle$ where T is the Euclidean unit tangent to $\mathscr{C}$. The Poincaré and Euclidean scalar products are related by $\langle A, B\rangle_F = 1/y^2\langle A, B\rangle$, and yT is the Poincaré unit tangent

T_F, so $k_g = -\langle T_F, \partial/\partial x \rangle_F$. But $\mathscr{C}$ is a Poincaré geodesic and $\partial/\partial x$ is clearly a Poincaré Killing vector field; hence, from the constant of motion lemma of Chapter 7 we have k_g constant along $\mathscr{C}$, i.e., $\mathscr{C}$ is a portion of a Euclidean circle. Since $\langle T_F, \partial/\partial x \rangle_F \neq 0$ at the highest point of $\mathscr{C}$, it must be that when $\mathscr{C}$ is vertical, i.e., $\langle T, \partial/\partial x \rangle = 0$, we must be at $y = 0$. We conclude the classical fact that the Poincaré geodesics are circular arcs (which can degenerate to vertical lines) that strike the x axis orthogonally.

The Deflection of Light

Consider the spatial path of a light ray emanating from a distant star that passes near the sun and strikes the earth, the earth again being considered as a test particle in the Schwarzschild field of the sun (Figure 8-2). As we have seen (Eqn. 5-18), the sectional curvatures of the Schwarzschild metric are of the order of $m_\odot/r^3$, with $m_\odot$ of the order 1.5×10^5 cm (see Misner, Thorne, and Wheeler, 1973, end papers, for a convenient table of constants) and r is very nearly the astronomical distance from the center of the sun. The radius of the sun is of the order 7×10^{10} cm. The sectional curvatures in the solar system are thus extremely small quantities and it is only when starlight passes very close to a massive body like the sun that there will be a detectable deflection.

An exact analysis of the deflection is given in most texts, and indeed such an exact analysis is needed when one considers extreme situations, as for example, deflection by a black hole (see, for example, Misner, Thorne, and Wheeler, Chapter 25). Here, however, I shall present only an approximate treatment and shall not attempt to estimate the errors made. I hope that these shortcomings are somewhat mitigated by the rather simple geometric picture.

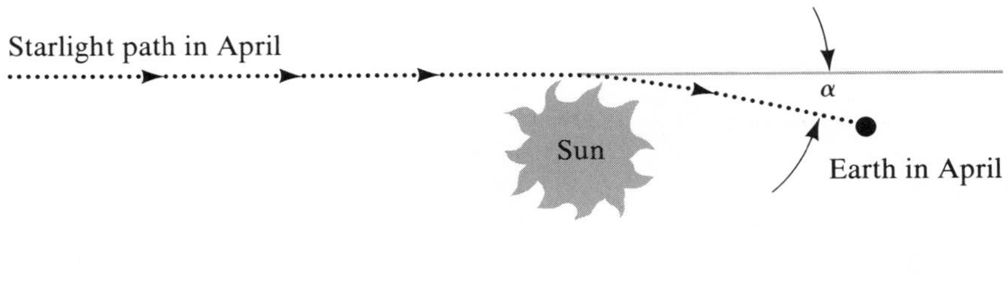

FIGURE 8-2

Let us compute the approximate Euclidean curvature of the starlight path as it passes near the sun. In order to use the lemma on page 93, we first write the Schwarzschild ds^2 in "isotropic" form, which exhibits the spatial metric dl^2 as conformally related to the flat Euclidean metric $d\tilde{l}^2 = d\rho^2 + \rho^2(d\theta^2 + \sin^2\theta\, d\varphi^2) = dx^2 + dy^2 + dz^2$ of R^3 (we wish to compare the spatial light path with a y-line, which is a geodesic in the flat metric):

$$ds^2 = -\frac{\left(1 - \frac{m}{2\rho}\right)^2}{\left(1 + \frac{m}{2\rho}\right)^2}\, dt^2 + \left(1 + \frac{m}{2\rho}\right)^4 d\tilde{l}^2$$

(see Adler, Bazin, and Schiffer, 1975, p. 198). The light ray traces out a geodesic in the Fermat metric for the spatial sections

$$dl_F^2 = f^2(\rho)\, d\tilde{l}^2$$

$$f(\rho) = \left(1 + \frac{m}{2\rho}\right)^3 \left(1 - \frac{m}{2\rho}\right)^{-1}.$$

We have written the Fermat metric as conformally related to the flat metric $d\tilde{l}$ of R^3.

The path of the light ray passing near the sun may be taken as being very close to the line $y = R$ in the flat xy plane with center of sun at the origin (Figure 8-3). Since the deflection is very small, we approximate the flat unit normal by $N \sim -\partial/\partial y$. Then from our lemma (applied to $dl_F = f \cdot d\tilde{l}$) we get for the flat-space curvature of the light ray path at radial coordinate ρ

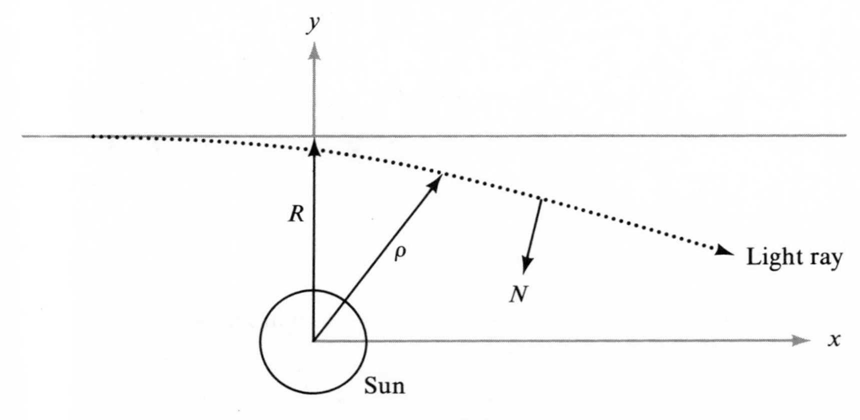

FIGURE 8-3

$$\begin{aligned} k_g &= N(\log f) \sim -\frac{\partial}{\partial y}(\log f) \\ &= \frac{m}{2\rho^2}\frac{y}{\rho}\left\{\frac{3}{1+\dfrac{m}{2\rho}} + \frac{1}{1-\dfrac{m}{2\rho}}\right\} \\ &\sim \frac{m}{2\rho^2}\cdot\frac{R}{\rho}\cdot 4 = \frac{2mR}{\rho^3}, \end{aligned}$$

where we have discarded terms involving m^2. Then, for the total angular change of the tangent vector to the light ray we get, since $y = R$ is a straight line of the flat metric,

$$\begin{aligned} \alpha &= \int_{\text{path}} k_g \, d\tilde{l} \sim \int_{-\infty}^{\infty} k_g \, dx \sim 2mR \int_{-\infty}^{\infty} \frac{dx}{\rho^3} \\ &\sim 2mR \int_{-\infty}^{\infty} \frac{dx}{(x^2 + R^2)^{3/2}} = \frac{4m}{R}, \end{aligned}$$

the classical expression obtained by Einstein in 1915.

Einstein's original 1911 calculation can be understood in the following light. Consider the path of a light ray in Minkowski space:

$$ds_0^2 = -dt^2 + dl_0{}^2$$

$$dl_0^2 = dx^2 + dy^2 + dz^2.$$

As mentioned in Chapter 2, in 1912 Einstein proposed that the speed of light $c(r)$ is related to the Newtonian potential $U = m/r$ by $c(r) = 1 - U = 1 - (m/r)$. By Fermat's principle, the spatial path of a light ray is a geodesic in the metric

$$\frac{dl_0}{c} = \frac{dl_0}{1 - \dfrac{m}{r}}$$

whose curvature, in the Euclidean metric dl_0 is approximately, from our lemma,

$$-\frac{\partial}{\partial y}\log\left(1 + \frac{m}{r}\right)^{-1} \sim \frac{my}{r^3} \sim \frac{mR}{r^3}.$$

This result, half the amount calculated for k_g above, leads to a total deflection of $2m/R$ rather than $4m/R$.

For an actual numerical calculation we need only a few well-known measurements. Recall that $m_\odot = \kappa M_\odot$. Let $R = r_\odot$ be the radius of the sun and let $R_\oplus$ be the mean distance of the earth from the sun. If we assume a circular orbit for the earth around the sun, then equating gravitational and centrifugal forces on the earth yields

$$\frac{m_\odot}{R^2_\oplus} = \frac{\kappa M_\odot}{R^2_\oplus} = \omega^2 R_\oplus,$$

where ω is the angular speed of the earth about the sun. Then

$$4\frac{m_\odot}{R} = 4\omega^2 R^2_\oplus \cdot \frac{R_\oplus}{R}.$$

Using

$$1 \text{ light second} = 3 \times 10^{10} \text{ cm}$$

$$R_\oplus = 500 \text{ light seconds}$$

$$R = 7 \times 10^{10} \text{ cm}$$

we see that

$$4\frac{m_\odot}{R} = \frac{4(2\pi)^2(500)^2(500)3}{(365)^2(24)^2(60)^4(7)} \text{ radians,}$$

or 1.75″ of arc.

9

Electromagnetism in Three-Space and Minkowski Space

Twisted Forms and the Vector Product

A *twisted* (exterior) differential form α^k on a manifold M^n assigns in a continuous fashion an ordinary exterior form to each *oriented* tangent space M_p in such a way that if the orientation of M_p is reversed, the form is replaced by its negative. (These forms are called, by de Rham, forms of *odd kind*; ordinary forms are then forms of *even kind*.) For example, the volume form ω^n on a (pseudo) Riemannian M^n assigns to the coordinate patch $(U; x^1, \ldots, x^n$, with $\partial/\partial x^1, \ldots, \partial/\partial x^n$ defining the positive orientation) the n-form

$$\omega_U^n = \sqrt{|g|}\, dx^1 \wedge \cdots \wedge dx^n.$$

If $e_1, \ldots, e_n$ is an orthonormal frame at $p \in U$, then $\omega_U^n(e_1, \ldots, e_n) = \pm 1$, the plus sign being used only if $e_1, \ldots, e_n$ defines the same orientation as $\partial/\partial x^1, \ldots, \partial/\partial x^n$.

One sees immediately that if X is a vector field and α^k is twisted, then $d\alpha^k$ and $i_X\alpha^k$ are again twisted. The map $X_p \to i_{X_p}\omega^n$ is an isomorphism of M_p^n onto the vector space of all *twisted* $(n - 1)$ forms at $p \in U$; consequently, every twisted 2-form on R^3 is of the form $i_X\omega^3$, where X is a vector field on R^3.

If M^n is orientable, then to each twisted form one can associate two ordinary forms; one need only specify a particular coordinate cover whose coordinate transition matrixes have positive Jacobians. Integration of twisted forms is discussed later in this chapter.

Recall the vector product $\mathbf{B} \times \mathbf{C}$ of a pair of vectors in R^3. Let ω^3 be the (twisted) volume form for R^3. If $\mathbf{B}$ and $\mathbf{C}$ are vectors, $\mathbf{B} \times \mathbf{C}$ is the unique *orientation dependent* vector such that

$$\mathbf{A} \cdot (\mathbf{B} \times \mathbf{C}) = \omega^3(\mathbf{A}, \mathbf{B}, \mathbf{C})$$

for all vectors $\mathbf{A}$. Clearly, the vector $\mathbf{B} \times \mathbf{C}$ reverses direction if the orientation of R^3 is reversed. $\mathbf{B} \times \mathbf{C}$ is thus not a true vector; it is usually called a *pseudo-vector* or *axial vector*.

E, B, and the (Heaviside-)Lorentz Force in Three-Space

The most basic fact of electromagnetism is that there is an invariant notion of the *charge* of a body. This charge is independent of the observer and of any metrical structure of space. The electromagnetic force on a test body of charge e units is given by the Heaviside-Lorentz force law

$$\mathbf{f}_L = e(\mathbf{E} + \mathbf{v} \times \mathbf{B}), \tag{9-1}$$

where $\mathbf{v}$ is the velocity of the test charge. This defines the fields $\mathbf{E}$ and $\mathbf{B}$, for the *electric field vector* $\mathbf{E}$ is determined by measuring the force on a charged body at rest and the *magnetic field vector* $\mathbf{B}$ is determined by measuring the force on moving charges. The formulation in Equation 9-1, while correct, has several drawbacks that point to a slightly different approach. The force $\mathbf{f}_L$ is a (true) vector and e is a scalar, hence $\mathbf{E}$ and $\mathbf{v} \times \mathbf{B}$ are vectors. $\mathbf{v}$ is definitely a vector, hence $\mathbf{v} \times \mathbf{B}$ can be a true vector only if $\mathbf{B}$ is a pseudo-vector; $\mathbf{B}$ *reverses direction when the orientation of* R^3 *is reversed!*

From our dictionary on page 63, $\mathbf{v} \times \mathbf{B}$ corresponds to the 1-form $-i_{\mathbf{v}}\mathcal{B}$, where $\mathscr{B}$ is the 2-form $\mathscr{B} = i_{\mathbf{B}}\omega^3$. Since ω^3 is twisted, and **B** is a pseudo-vector, $\mathscr{B}$ is an ordinary 2-form. In terms of any coordinate system

$$\mathbf{B} = B^1 \frac{\partial}{\partial x^1} + B^2 \frac{\partial}{\partial x^2} + B^3 \frac{\partial}{\partial x^3},$$

and, from Equations 6-3 and 6-4,

$$\begin{aligned}\mathscr{B} &= i_{\mathbf{B}} \cdot \sqrt{g}\, dx^1 \wedge dx^2 \wedge dx^3 \\ &= \sqrt{g}\,(B^1\, dx^2 \wedge dx^3 + B^2\, dx^3 \wedge dx^1 + B^3\, dx^1 \wedge dx^2),\end{aligned}$$

i.e., $B_{23} = \sqrt{g}\, B^1, B_{31} = \sqrt{g}\, B^2, B_{12} = \sqrt{g}\, B^3$. Since $\mathscr{B}^2$ is independent of orientation, we see that the magnetic field is very conveniently represented by an exterior 2-form $\mathscr{B}^2$ in 3-space.

Using the metric of 3-space we can associate to the vector **E** the covector (1-form)

$$\mathscr{E} = E_1\, dx^1 + E_2\, dx^2 + E_3\, dx^3,$$

defined as usual by $\mathscr{E}(\mathbf{A}) = \mathbf{E} \cdot \mathbf{A}$, for all vectors **A**. If we let $\mathfrak{f}_L$ be the covector associated in the same fashion to $\mathbf{f}_L$, we may write Equation 9-1 in a covariant form *independent of orientation:*

$$\mathfrak{f}_L = e(\mathscr{E} - i_{\mathbf{v}}\mathscr{B}). \tag{9-2}$$

It is natural to use this covariant form since both the Lorentz force and the electric field enter naturally in line integrals over *oriented* paths. Thus

$$\int_C \mathfrak{f}_L = \int_C \mathbf{f}_L \cdot d\mathbf{r}$$

is the work done by the force, and

$$\int_C \mathscr{E} = \int_C \mathbf{E} \cdot d\mathbf{r}$$

is the "voltage." Also, the magnetic field enters via surface integrals over *oriented* surfaces. Thus

$$\iint_{\Sigma_2} \mathscr{B} = \iint_{\Sigma_2} \mathbf{B} \cdot \mathbf{N}\, dA$$

is the "magnetic flux" through the surface Σ_2.

Electromagnetism in Minkowski Space

It is clear from Equation 9-1 that the definitions of **E** and **B** depend classically on the motion of the observer (a test particle at rest with respect to one inertial system will be in motion with respect to another). Minkowski gave a remarkable synthesis of **E** and **B** (based mainly on previous considerations of Einstein) by considering the electromagnetic field in space-time, i.e., Minkowski space.

Consider an inertial system. To the classical force $\mathbf{f}_C = \mathbf{f}_L$ one associates the Minkowski 4-force (see Eqn. 1-11)

$$f = \gamma\left(\mathbf{f}_L \cdot \mathbf{v}\frac{\partial}{\partial t} + \mathbf{f}_L\right).$$

Using the Minkowski metric $ds^2 = -dt^2 + dx^2 + dy^2 + dz^2$, we have the associated covector $f_i = g_{ij} f^j$, i.e.,

$$\begin{aligned} \mathfrak{f} &= \gamma(-\mathbf{f}_L \cdot \mathbf{v}\, dt + \mathfrak{f}_L) \\ &= \gamma(-i_{\mathbf{v}}(\mathfrak{f}_L)\, dt + \mathfrak{f}_L). \end{aligned}$$

From Equation 9-2, $i_{\mathbf{v}}(\mathfrak{f}_L) = e\, i_{\mathbf{v}}(\mathscr{E})$, since $i_{\mathbf{v}}(i_{\mathbf{v}}\mathscr{B}) = 0$. Consequently,

$$\mathfrak{f} = -e\gamma(i_{\mathbf{v}}(\mathscr{E})\, dt - \mathscr{E} + i_{\mathbf{v}}\mathscr{B}).$$

Introducing the unit velocity 4-vector $u = \gamma(\partial/\partial t + \mathbf{v})$ (as in Eqn. 1-9) we see that in the electromagnetic case $\mathfrak{f}$ is given by

$$\begin{cases} \mathfrak{f} = -e i_u \mathscr{F} \\ \mathscr{F} = \mathscr{E} \wedge dt + \mathscr{B} \end{cases} \tag{9-3}$$

We have shown that the intrinsic 4-velocity u along the world line and the intrinsic 4-force $\mathfrak{f}$ along the world line are *linearly related* by means of a 2-form $\mathscr{F}$. $\mathscr{F}$ is thus *intrinsically defined,* in spite of its initial coordinate expression $\mathscr{E} \wedge dt + \mathscr{B}$.

In terms of inertial coordinates,

$$\begin{aligned} \mathscr{F} = {} & (E_1\, dx^1 + E_2\, dx^2 + E_3\, dx^3) \wedge dt \\ & + B_1\, dx^2 \wedge dx^3 + B_2\, dx^3 \wedge dx^1 + B_3\, dx^1 \wedge dx^2. \end{aligned}$$

Writing $\mathscr{F} = F_{i<j}\, dx^i \wedge dx^j$, we have

$$(F_{ij}) = \begin{pmatrix} 0 & -E_1 & -E_2 & -E_3 \\ E_1 & 0 & B_3 & -B_2 \\ E_2 & -B_3 & 0 & B_1 \\ E_3 & B_2 & -B_1 & 0 \end{pmatrix} \tag{9-4}$$

and Equation 9-3 becomes

$$f_i = -eu^j F_{ji} = eF_{ij}u^j, \tag{9-5}$$

where (F_{ij}) is the electromagnetic field tensor, introduced by Minkowski. The form $\mathcal{F} = \mathcal{E} \wedge dt + \mathcal{B}$ was apparently first introduced by Hargreaves.

The intrinsic form $\mathcal{F}$ takes the particular form $\mathcal{E} \wedge dt + \mathcal{B}$ in an inertial coordinate system (t, x, y, z). Let (t', x', y', z') be another inertial system, the two systems being related by a Lorentz transformation of the type in Equation 1-2. Then $\mathcal{F} = \mathcal{E} \wedge dt + \mathcal{B} = \mathcal{E}' \wedge dt' + \mathcal{B}'$ yields rather immediately the famous transformation of field components

$$\begin{cases} E_x = E'_x & B_x = B'_x \\ E_y = \gamma(E'_y + vB'_z) & B_y = \gamma(B'_y - vE'_z) \\ E_z = \gamma(E'_z - vB'_y) & B_z = \gamma(B'_z + vE'_y) \end{cases}$$

Integration of Twisted Forms

Let M^n be a manifold and let α^k be an ordinary exterior k-form on M^n. Ordinary k-forms are integrated over oriented simplexes; if $\varphi: \Delta_k \to M^n$ is a smooth map of an *oriented* Euclidean k-simplex Δ_k, then the integral of α over $\varphi(\Delta_k)$ is defined in the usual way by means of the pull back

$$\int_{\varphi(\Delta_k)} \alpha^k = \int_{\Delta_k} \varphi^* \alpha^k.$$

Twisted k-forms β^k are integrated not over oriented simplexes but rather simplexes arising from *oriented maps*. (Since we shall be concerned only with singular simplexes that are diffeomorphic images of Euclidean simplexes lying entirely in coordinate patches, we shall avail ourselves of the following simplified prescription; see de Rham, 1955, for the general treatment). An oriented map $\varphi: |\Delta_k| \to M^n$ of an unoriented Euclidean simplex $|\Delta_k|$ assigns to each orientation of $|\Delta_k|$ coherent orientations of each coordinate patch holding $\varphi(|\Delta_k|)$ in such a way that reversing the orientation of $|\Delta_k|$ reverses the orientation of each such patch. The integral of a *twisted* β^k on M^n over $\varphi(|\Delta_k|)$ is then defined as follows. Choose an orientation of $|\Delta_k|$, and call Δ_k the resulting oriented simplex. φ induces an orientation of a coordinate patch U holding $\varphi(|\Delta_k|)$ and this orientation of U picks out a unique ordinary form β_U^k in U. Define

$$\int_{\varphi(|\Delta_k|)} \beta^{k\cdot} = \int_{\varphi(\Delta_k)} \beta_U^k = \int_{\Delta_k} \varphi^* \beta_U^k.$$

It is clear that this definition is in fact independent of the orientation of $|\Delta_k|$ chosen.

Oriented maps will frequently arise as follows. Let $\varphi: |\Delta_k| \to M^n$ be an embedding of the unoriented Euclidean simplex. Then φ can be oriented if one can orient the normal space to $\varphi(|\Delta_k|)$. Suppose, for instance, that there is a continuous distribution of $(n - k)$ linearly independent vector fields $N_1, \ldots, N_{n-k}$ defined along $\varphi(|\Delta_k|)$ that are transverse to $\varphi(|\Delta_k|)$. Then if $Y_1, \ldots, Y_k$ are vector fields on $|\Delta_k|$ defining an orientation of $|\Delta_k|$, we let $N_1, \ldots, N_{n-k}, \dot{\varphi} Y_1, \ldots, \dot{\varphi} Y_k$ define an orientation of a neighborhood of $\varphi(|\Delta_k|)$. Twisted forms in physics are usually integrated over such "framed" simplexes, or simplexes with "transverse" orientation.

Twisted forms can be integrated over "chains," i.e., formal sums of simplexes arising from oriented maps.

Note that an embedding of an n-simplex into M^n is automatically an oriented map. A twisted n-form (e.g., a volume form) can be integrated over any triangulation (oriented or not) of a compact portion of M^n. More generally, by means of a partition of unity, one can integrate twisted n-forms over a compact domain with piecewise regular boundary (see Warner, 1971, for this procedure in the oriented case).

Stokes's theorem holds for twisted forms and simplexes arising from oriented maps since an oriented map of a simplex yields an oriented map of its boundary. Note that if $N_1, \ldots, N_{n-k}$ is a transverse frame to $\varphi(|\Delta_k|)$ and if N is the outward normal to the boundary of $\varphi(|\Delta_k|)$ that is tangent to $\varphi(|\Delta_k|)$, then $N_1, \ldots, N_{n-k}, N$ defines a transverse frame for the boundary.

If V^{n-1} is an embedded submanifold of the (pseudo) Riemannian M^n and if N is any local unit normal to V^{n-1}, then

$$\omega_V^{n-1} = i_N \omega^n$$

is a well-defined $(n - 1)$-twisted volume form for V. (If the local tangent vectors $Y_1, \ldots, Y_{n-1}$ define an orientation of V^{n-1}, then $N, Y_1, \ldots, Y_{n-1}$ defines an orientation of a neighborhood of V^{n-1} in M^n, and the twisted form ω_V^{n-1} is independent of the choice of local normal N.)

The Charge-Current Three-Form in Minkowski Space

Minkowski space is endowed with its light-cone structure. We shall assume that Minkowski space is *time-oriented,* i.e., a continuous selection of "forward" light cones has been made. Thus, to each spatial hypersurface there is a distinguished normal that points "forward" in time.

The light-cone structure of Minkowski space defines a conformal structure, for if ds^2 is the Minkowski metric, and if $d\tilde{s}^2$ is any other pseudo-Riemannian metric that has at each point the same light-cone structure, then $d\tilde{s}^2 = f\, ds^2$, where f is some positive smooth function. This light-cone or conformal structure is more primitive than the Minkowski metric.

We shall assume the following.

Axiom of charge There exists a twisted 3-form $\mathscr{S}$ the "charge-current" 3-form in Minkowski space, dependent only on the conformal structure, such that if V^3 is any compact regular domain of a space-like hypersurface,

$$Q(V^3) = \int_{V^3} \mathscr{S}$$

is the charge contained in V^3.

The forward direction of time is used to orient the normal space to V^3.

In terms of an inertial coordinate system (t, x, y, z), $\mathscr{S}$ has the unique expression

$$\begin{cases} \mathscr{S} = \sigma dx \wedge dy \wedge dz - \mathscr{J} \wedge dt \\ \mathscr{J} = j^1\, dy \wedge dz + j^2\, dz \wedge dx + j^3\, dx \wedge dy. \end{cases} \tag{9-6}$$

Here σ is the *charge density* and $\mathscr{J}$ is the *current 2-form.* In terms of the Minkowski volume 4-form $\omega^4 = dt \wedge dx \wedge dy \wedge dz$, we have

$$\mathscr{S} = i_J \omega^4 \tag{9-7}$$

where

$$\begin{cases} J = \sigma \dfrac{\partial}{\partial t} + \mathbf{j} \\ \mathbf{j} = j^1 \dfrac{\partial}{\partial x} + j^2 \dfrac{\partial}{\partial y} + j^3 \dfrac{\partial}{\partial z}. \end{cases} \tag{9-8}$$

J is the *current 4-vector.* Note that while $\mathcal{S}$ is assumed independent of the metric (i.e., dependent only on the conformal structure), J is linked to the metric via the volume form ω^4. When V_t^3 is a spatial hypersurface defined by t = constant,

$$Q = \int_{V_t^3} \mathcal{S}^3 = \int_{V_t^3} \sigma dx \wedge dy \wedge dz = \int_{V_t^3} \sigma \omega_V^3.$$

It is not difficult to see that $\mathcal{S}$ (and therefore J) is determined if we know σ for each spatial section, i.e., for the spatial section of each inertial observer. Thus charge alone determines $\mathcal{S}$.

The Hodge $*$-Operator

Since we shall require the $*$-operator in both three and four dimensions, we shall review the theory in any (pseudo)-Riemannian M^n.

Let $I = i_1, \ldots, i_k$, $J = j_1, \ldots, j_l$, etc., be multi-indexes. Locally, a k-form of even (ordinary) or odd (twisted) type on a (pseudo)-Riemannian M^n has a coordinate expression

$$\alpha^k = a_{\underset{\rightarrow}{I}}\, dx^I = a_{i_1 < \cdots < i_k} dx^{i_1} \wedge \ldots \wedge dx^{i_k},$$

where $\underset{\rightarrow}{I}$ indicates that indexes are in increasing order, $i_1 < \cdots < i_k$, during the summation. The *dual* form $*\alpha^k$ is defined to be the $(n - k)$-form of *opposite type*

$$\begin{cases} *\alpha = a^*{}_{\underset{\rightarrow}{J}}\, dx^J \\ a_J^* = \sqrt{|g|}\, a^L \epsilon_{\underset{\rightarrow}{L} J}, \end{cases} \tag{9-9}$$

that is,

$$a^*_{j_1 \cdots j_{n-k}} = \sqrt{|g|}\, a^{l_1 \cdots l_k} \epsilon_{l_1 < \cdots < l_k\, j_1 \cdots j_{n-k}},$$

where ϵ is the classical permutation symbol

$$\epsilon_{\alpha_1 \cdots \alpha_n} = \begin{cases} +1 \text{ if } \alpha_1 \cdots \alpha_n \text{ is an even permutation of } 1, \ldots, n \\ -1 \text{ if } \alpha_1 \cdots \alpha_n \text{ is an odd permutation of } 1, \ldots, n \\ 0 \text{ otherwise.} \end{cases}$$

Note that for $J = j_{1,\ldots,}j_{n-k}$ fixed there is at most one nonvanishing term in the expression for a^*_J since LJ must be a permutation of $1, \ldots, n$ and $l_1 < \cdots < l_k$.

For example, in 3-space with coordinates x^1, x^2, x^3, if $\mathscr{E} = E_1\, dx^1 + E_2\, dx^2 + E_3\, dx^3$ is the electric field 1-form (of even kind), then

$$\begin{aligned} *\mathscr{E} &= E^*_{j<k}\, dx^j \wedge dx^k = \sqrt{g}\, E^i \epsilon_{ij<k}\, dx^j \wedge dx^k \\ &= \sqrt{g}\, (E^1\, dx^2 \wedge dx^3 + E^2\, dx^3 \wedge dx^1 + E^3\, dx^1 \wedge dx^2), \end{aligned}$$

that is,

$$*\mathscr{E} = i_{\mathbf{E}}\omega^3 \equiv \mathscr{D}. \tag{9-10}$$

$\mathscr{D} \equiv *\mathscr{E}$ is, in the case of a "noninductive medium," the *displacement* 2-form (a twisted form).

Note, in general, that if X is a vector on M^n, $X = X^i \partial/\partial x^i$, and if $\mathscr{X} = X_i\, dx^i$ is the covariant version of X, then

$$\begin{aligned} i_X\omega^n &= i_X \sqrt{|g|}\, dx^1 \wedge \ldots \wedge dx^n \\ &= i_X \sqrt{|g|}\, \epsilon_{i_1<\cdots<i_n} dx^{i_1} \wedge \cdots \wedge dx^{i_n} \end{aligned}$$

which by Equation 6-5 yields

$$i_X\omega^n = \sqrt{|g|}\, X^j \epsilon_{ji_2<\cdots<i_n}\, dx^{i_2} \wedge \cdots \wedge dx^{i_n},$$

that is,

$$i_X\omega^n = *\mathscr{X}. \tag{9-11}$$

Note further that if X is a vector (resp. pseudo-vector) then $*\mathscr{X}$ is an $(n-1)$-form of odd (resp. even) kind. Thus if we let $\mathscr{H}$ be the covariant version of the magnetic field pseudo-vector **B**

$$\mathscr{H} \equiv B_1\, dx^1 + B_2\, dx^2 + B_3\, dx^3 \tag{9-12}$$

then $\mathscr{B} = i_{\mathbf{B}}\omega^3 = *\mathscr{H}$. $\mathscr{H}$ is a 1-form of odd kind.

Returning to a general M^n, if $\alpha^k = a_{\underset{\rightarrow}{I}}\, dx^I$ and $\beta^k = b_{\underset{\rightarrow}{I}}\, dx^I$ are both k-forms of the same kind, then $\alpha \wedge *\beta$ is a multiple of the volume form. For coefficient, note

$$\begin{aligned}\alpha \wedge *\beta &= a_{\underset{\to}{I}} b^*_{\underset{\to}{J}}\, dx^I \wedge dx^J \\ &= a_{\underset{\to}{I}} \sqrt{|g|}\, \epsilon_{\underset{\to}{L}\,\underset{\to}{J}} b^L\, dx^I \wedge dx^J \\ &= a_{\underset{\to}{I}} \sqrt{|g|}\, \epsilon_{\underset{\to}{L_J}\,\underset{\to}{J}} b^{L_J} dx^I \wedge dx^J,\end{aligned}$$

where $\underset{\to}{L_J} = l_1 < \cdots < l_k$ is the unique multi-index complementary to $\underset{\to}{J} = j_1 < \cdots < j_{n-k}$. Thus

$$\begin{aligned}\alpha \wedge *\beta &= a_{\underset{\to}{L_J}} b^{L_J} \sqrt{|g|}\, \epsilon_{\underset{\to}{L_J}\,\underset{\to}{J}}\, dx^{L_J} \wedge dx^J \\ &= a_{\underset{\to}{L_J}} b^{L_J} \omega^n \qquad (9\text{-}13) \\ &= a_{\underset{\to}{I}} b^I \omega^n = \langle \alpha, \beta \rangle \omega^n,\end{aligned}$$

where $\langle \alpha, \beta \rangle = a_{i_1<\cdots<i_k} b^{i_1 \cdots i_k}$. In particular $\langle \alpha, \alpha \rangle = \|\alpha\|^2$ is the norm-squared of α (this can be negative in a pseudo-Riemannian M^n). Also note that for the 0-form 1 (the constant function)

$$\begin{aligned}*1 &= 1^*_{i_1<\cdots<i_n}\, dx^{i_1} \wedge \cdots \wedge dx^{i_n} \\ &= \sqrt{|g|}\, dx^1 \wedge \cdots \wedge dx^n \\ &= \omega^n.\end{aligned}$$

The Laws of Gauss and Ampère–Maxwell

Consider the Hodge $*$-operator in Minkowski space. The form $\mathcal{F} = \mathcal{E} \wedge dt + \mathcal{B}$, an ordinary 2-form, has been defined via the *intensity* of the electromagnetic field, i.e., via the Lorentz force. We now construct the dual twisted 2-form $*\mathcal{F}$ and relate it to the distribution of charge. (We shall retain the factor $\sqrt{|g|} = 1$ for future reference.) From Equation 9-9,

$$*\mathcal{F} = F^*_{k<l}\, dx^k \wedge dx^l$$

$$F^*_{kl} = \sqrt{|g|}\, F^{i<j} \epsilon_{ijkl}.$$

Using the metric tensor $(g_{ij}) = \operatorname{diag}(-1, 1, 1, 1)$ of Minkowski space, and $F_{\alpha 0} = E_\alpha$, $F_{\alpha\beta} = B_{\alpha\beta}$, $B_{23} = B_1$, etc., we have

$$
\begin{aligned}
(F^*{}_{ij}) &= \sqrt{|g|}\begin{pmatrix} 0 & F^{23} & F^{31} & F^{12} \\ -F^{23} & 0 & F^{03} & F^{20} \\ -F^{31} & -F^{03} & 0 & F^{01} \\ -F^{12} & -F^{20} & -F^{01} & 0 \end{pmatrix} \\
&= \begin{pmatrix} 0 & B_1 & B_2 & B_3 \\ -B_1 & 0 & E_3 & -E_2 \\ -B_2 & -E_3 & 0 & E_1 \\ -B_3 & E_2 & -E_1 & 0 \end{pmatrix}
\end{aligned}
\tag{9-14}
$$

Thus we can write

$$*\mathscr{F} = -\mathscr{H} \wedge dt + \mathscr{D}, \tag{9-15}$$

where again

$$
\begin{cases}
\mathscr{H} = B_1\, dx^1 + B_2\, dx^2 + B_3\, dx^3 \\
\mathscr{D} = E_1\, dx^2 \wedge dx^3 + E_2\, dx^3 \wedge dx^1 + E_3\, dx^1 \wedge dx^2.
\end{cases}
$$

We now postulate how $\mathscr{F}$ is related to charge.

Axiom of Gauss In all inertial coordinate systems

$$\operatorname{div} \mathbf{E} = 4\pi\sigma.$$

Theorem of Ampere–Maxwell $d * \mathscr{F} = 4\pi\mathscr{S}$ and consequently

$$\operatorname{curl} \mathbf{B} = 4\pi\mathbf{j} + \frac{\partial \mathbf{E}}{\partial t}\cdot$$

PROOF In a given inertial coordinate system of Minkowski space let $\mathbf{d} = dx \wedge \partial/\partial x + dy \wedge \partial/\partial y + dz \wedge \partial/\partial z$ be the spatial exterior derivative and $d = dt \wedge \partial/\partial t + \mathbf{d}$ the exterior derivative for the full Minkowski space. Then

$$d{*}\mathscr{F} = d(-\mathscr{H} \wedge dt + \mathscr{D}) = \left(-\mathbf{d}\mathscr{H} + \frac{\partial \mathscr{D}}{\partial t}\right) \wedge dt + \mathbf{d}\mathscr{D},$$

where

$$\frac{\partial}{\partial t} D_{\alpha<\beta}\, dx^\alpha \wedge dx^\beta = \frac{\partial D_{\alpha<\beta}}{\partial t}\, dx^\alpha \wedge dx^\beta.$$

By Gauss's law

$$\mathbf{d}\mathcal{D} = \mathbf{d}(E_1\, dx^2 \wedge dx^3 + E_2\, dx^3 \wedge dx^1 + E_3\, dx^1 \wedge dx^2)$$
$$= \operatorname{div} \mathbf{E}\, dx \wedge dy \wedge dz = 4\pi\sigma\, dx \wedge dy \wedge dz .$$

Consequently, using Equation 9-6,

$$d * \mathcal{F} - 4\pi\mathcal{S}$$
$$= \left[-\mathbf{d}\mathcal{H} + \frac{\partial \mathcal{D}}{\partial t} + 4\pi(j^1\, dx^2 \wedge dx^3 + j^2\, dx^3 \wedge dx^1 + j^3\, dx^1 \wedge dx^2)\right] \wedge dt$$

has the property that its restriction to the spatial section $dt = 0$ vanishes. Since this must be true for the spatial sections of *every* inertial coordinate system, it is not difficult to see that $d * \mathcal{F} - 4\pi\mathcal{S}$ must itself vanish. Consequently,

$$\mathbf{d}\mathcal{H} = 4\pi\mathcal{J} + \frac{\partial \mathcal{D}}{\partial t},$$

which is the form version of Ampère's law curl $\mathbf{B} = 4\pi\mathbf{j}$ augmented by Maxwell's "displacement current" $\partial\mathbf{E}/\partial t$.

It is instructive to see why $\mathcal{D}$ and $\mathcal{H}$ are *twisted* forms. The integral form of Gauss's law states that if Σ_2 is a closed surface in 3-space bounding a region $\mathcal{R}_3$, and if $\mathbf{N}$ is the outward pointing normal to Σ_2, then (see p. 104ff) Σ_2 together with $\mathbf{N}$ forms a suitable object over which to integrate a twisted 2-form, and

$$\int_{(\Sigma_2,\, \mathbf{N})} \mathcal{D} = \int_{\mathcal{R}_3} d\mathcal{D} = 4\pi \int_{\mathcal{R}_3} \sigma\omega^3 = 4\pi Q(\mathcal{R}_3).$$

In terms of vector analysis

$$\int_{(\Sigma_2,\, \mathbf{N})} \mathcal{D} = \int_{\partial\mathcal{R}_3} i_{\mathbf{E}}\omega^3 = \int_{\partial\mathcal{R}_3} \mathbf{E} \cdot \mathbf{N} dA ,$$

and since $\mathbf{N}$ is the outward normal to Σ_2 this last integral clearly makes sense without any consideration of orientation of $\mathcal{R}_3$ or of Σ_2.

Similarly, let Σ_2 be a two-sided piece of surface with boundary curve $\mathcal{C}_1 = \partial\Sigma_2$. To pick out one of the two possible directions of transit across Σ_2 we choose one of the two possible unit normal fields $\mathbf{N}$ to Σ_2 (Figure 9-1). Note that in this process of orienting the transversal to Σ_2 we

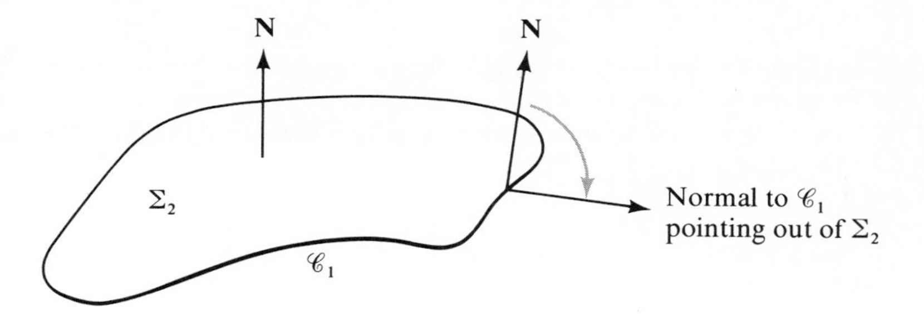

FIGURE 9-1

automatically orient the two-dimensional transversal to $\mathscr{C}_1$. Ampère's law states

$$\int_{\mathscr{C}_1} \mathscr{H} = \int_{(\Sigma_2,\mathbf{N})} d\mathscr{H} = 4\pi \int_{(\Sigma_2,\mathbf{N})} \mathscr{J}$$

and we shall see shortly that $\int_{(\Sigma_2,\mathbf{N})}\mathscr{J}$ is the amount of current flowing through Σ_2 in the given mode of transit **N**. (Ampère's law was corrected by Maxwell by the addition of his displacement current, but we should note that the correct law, as given in the theorem, follows directly from Gauss's law and special relativity.)

$\mathscr{D}$ and $\mathscr{H}$ lead to quantities of charge and current. On the other hand, $\mathscr{E}$ and $\mathscr{B}$ lead to intensities or field strengths. It is a guiding principle that forms of even kind measure *intensities* while twisted forms measure *quantities*. This is exemplified in the use of a twisted form $\mathscr{S}$ to measure quantity of charge.

The following is a classic consequence of Ampère-Maxwell.

Corollary Charge is conserved.

PROOF From $d{*}\mathscr{F} = 4\pi\mathscr{S}$ and $d^2 = 0$, we have

$$0 = d\mathscr{S} = di_J\omega^4 = \mathscr{L}_J\omega^4 = (\operatorname{div} J)\omega^4.$$

This can be interpreted as conservation of charge in two ways. First, let $\mathscr{U}_0$ be a compact subset (for example, a 3-ball) of the spatial section $t = 0$ of some inertial coordinate system, and let $\mathscr{U}_\varepsilon$ be the image of $\mathscr{U}_0$ under the flow $\{\varphi_\varepsilon\}$ in M^4 generated by the vector field J. Then the charge $Q(\varepsilon) = Q(\mathscr{U}_\varepsilon)$ in $\mathscr{U}_\varepsilon$ satisfies, from Equation 6-9,

$$\frac{dQ}{d\varepsilon} = \frac{d}{d\varepsilon}\int_{\mathscr{U}_\varepsilon} \mathscr{S} = \int_{\mathscr{U}_\varepsilon} \mathscr{L}_J\mathscr{S} = \int_{\mathscr{U}_\varepsilon} i_J d\mathscr{S} + di_J\mathscr{S} = 0.$$

Thus the flow generated by J leaves the charge invariant. Second, let $\mathcal{U}_0$ again be given but keep this set spatially fixed, i.e., let $\tilde{\mathcal{U}}_t$ be the image of $\mathcal{U}_0$ under the time translation for a given inertial system. Then for $\tilde{Q}(t) = \tilde{Q}(\tilde{\mathcal{U}}_t)$ we have

$$\frac{d\tilde{Q}}{dt} = \frac{d}{dt}\int_{\tilde{\mathcal{U}}_t} \mathcal{S} = \int_{\tilde{\mathcal{U}}_t} \mathcal{L}_{\partial/\partial t}\mathcal{S} = \int_{\tilde{\mathcal{U}}_t} di_{\partial/\partial t}\mathcal{S} = \int_{\partial\tilde{\mathcal{U}}_t} i_{\partial/\partial t}\mathcal{S}$$

$$= \int_{\partial\tilde{\mathcal{U}}_t} i_{\partial/\partial t} i_J \omega^4 = -\int_{\partial\tilde{\mathcal{U}}_t} i_J\omega^3 = -\int_{\partial\tilde{\mathcal{U}}_t} i_{\mathbf{j}}\omega^3,$$

where $\omega^3 = i_{\partial/\partial t}\omega^4$ is the 3-volume form for the spatial sections and $\mathbf{j}$ is defined in Equation 9-8. Since charge is conserved (in the sense of our first computation above) we see that

$$\int_{\partial\tilde{\mathcal{U}}_t} i_{\mathbf{j}}\omega^3$$

represents the net rate at which charge is leaving $\tilde{\mathcal{U}}_t$; $\mathbf{j}$ is thus the current vector in 3-space (see Eqn. 6-12).

In all that follows we shall only be concerned with *convective currents,* i.e., where the current 4-vector J is a multiple of the velocity 4-vector of a fluid $J = \tilde{\sigma}u$, $\tilde{\sigma}$ being an invariant i.e., independent of coordinates. To see what $\tilde{\sigma}$ is, note $\mathcal{S} = i_J\omega^4 = \tilde{\sigma}i_u dt \wedge dx^1 \wedge dx^2 \wedge dx^3$ and $u = \gamma(\partial/\partial t + \mathbf{v})$ as in Equation 1-8. Thus

$$\mathcal{S} = \tilde{\sigma}\gamma\, dx^1 \wedge dx^2 \wedge dx^3 - \tilde{\sigma}\gamma(v^1\, dx^2 \wedge dx^3 + v^2\, dx^3 \wedge dx^1 + v^3\, dx^1 \wedge dx^2) \wedge dt$$

is to be compared with $\mathcal{S} = \sigma\, dx^1 \wedge dx^2 \wedge dx^3 - \mathcal{J} \wedge dt$, and so $\sigma = \tilde{\sigma}\gamma$. If an "infinitesimal" region has rest volume vol_0 (the volume when seen from a co-moving observer) its volume as seen from the inertial system (t, x^1, x^2, x^3) will be $\mathsf{vol} = \gamma^{-1}\mathsf{vol}_0$, because of the Lorentz contraction in the one direction of motion. Thus, if we let σ_0 be the *rest density of charge,* $\sigma_0\mathsf{vol}_0 = \sigma\ \mathsf{vol} = \sigma\gamma^{-1}\mathsf{vol}_0$ or $\sigma_0 = \tilde{\sigma}$. Finally, then

$$J = \sigma_0 u \tag{9-16}$$

and also $\mathbf{j} = \sigma_0\gamma\mathbf{v} = \sigma\mathbf{v}$.

Faraday's Law and the Absence of Magnetic Monopoles

Gauss's law $\mathsf{div}\ \mathbf{E} = 4\pi\sigma$ shows that charges are a source of the electric field $\mathbf{E}$. The Ampère-Maxwell law $\mathsf{curl}\ \mathbf{B} = 4\pi\mathbf{j} + \partial\mathbf{E}/\partial t$ shows that currents and changing electric fields are sources of the magnetic field **B**. The next axiom states that these are the *only* sources of the magnetic field, that is, there is no magnetic analog of the electric charge.

Axiom of no magnetic charges In all inertial systems div **B** = 0.

Theorem of Faraday $d\mathcal{F} = 0$, and consequently curl $\mathbf{E} = -\partial \mathbf{B}/\partial t$.

PROOF Using the same notation as in the proof of the Ampère-Maxwell law, we have

$$d\mathcal{F} = d(\mathcal{E} \wedge dt + \mathcal{B}) = \mathbf{d}\mathcal{E} \wedge dt + dt \wedge \frac{\partial \mathcal{B}}{\partial t} + \mathbf{d}\mathcal{B}$$

and

$$\mathbf{d}\mathcal{B} = \text{div}\, \mathbf{B}\, dx \wedge dy \wedge dz = 0.$$

Consequently $d\mathcal{F} = \left(\mathbf{d}\mathcal{E} + \frac{\partial \mathcal{B}}{\partial t}\right) \wedge dt$ restricts to the zero 3-form on each spatial section. If then follows again that $d\mathcal{F} = 0$, and

$$\mathbf{d}\mathcal{E} = -\frac{\partial \mathcal{B}}{\partial t},$$

which is the form version of Faraday's law.

Consider the integral version of Faraday's law. Let $\Sigma_2(t)$ be a compact *oriented* surface that is perhaps moving in 3-space. From Equation 6-14, where **v** is the velocity of motion of the surface,

$$\begin{aligned}\frac{d}{dt}\int_{\Sigma_2(t)} \mathcal{B} &= \int_{\Sigma_2(t)} \frac{\partial \mathcal{B}}{\partial t} + i_{\mathbf{v}}\, \mathbf{d}\mathcal{B} + \mathbf{d}i_{\mathbf{v}}\mathcal{B} \\ &= \int_{\Sigma_2(t)} -\mathbf{d}\mathcal{E} + \mathbf{d}i_{\mathbf{v}}\mathcal{B} = -\int_{\partial\Sigma_2(t)} (\mathcal{E} - i_{\mathbf{v}}\mathcal{B}).\end{aligned}$$

In vector analysis terms, choose an orientation of R^3. Then **B** is given and also the orientation of $\Sigma_2(t)$ will determine a unit normal **N** to $\Sigma_2(t)$.

$$\frac{d}{dt}\int_{\Sigma_2(t)} \mathcal{B} = \frac{d}{dt}\int_{\Sigma_2(t)} \mathbf{B}\cdot\mathbf{N}\, dA = -\int_{\partial\Sigma_2(t)} (\mathbf{E} + \mathbf{v}\times\mathbf{B})\cdot d\mathbf{r}.$$

In particular, if the surface Σ_2 is fixed in space,

$$\int_{\partial \Sigma_2} \mathbf{E} \cdot d\mathbf{r} = -\frac{d}{dt} \int_{\Sigma_2} \mathbf{B} \cdot \mathbf{N} \, dA$$

gives the "voltage drop" around $\partial\Sigma_2$ as the negative time derivative of the "magnetic flux" through Σ_2 (Faraday's original formulation).

10

Electromagnetism in General Relativity

Maxwell's Equations

Let M^4 be the space-time of general relativity. A pure electromagnetic field is described by an exterior 2-form (of even kind) $\mathscr{F}$; in local coordinates $\mathscr{F} = \mathscr{E} \wedge dt + \mathscr{B}$, where $\mathscr{B}$ contains no dt terms. Given a test particle of charge e and unit 4-velocity u, the (Heaviside)-Lorentz force law will again be taken to be the contravariant version of Equation 9-5:

$$f^i = eF^i{}_j u^j.$$

If m_0 is the rest mass of the test particle, this becomes

$$m_0(\nabla_u u)^i = eF^i{}_j u^j, \tag{10-1}$$

and this formula again serves to define operationally the electromagnetic field tensor $(F^i{}_j)$.

In local coordinates $t = x^0, x^1, x^2, x^3, F_{ij} = g_{ik}F^k{}_j$ has the same matrix (Eqn. 9-4) as in Minkowski space.

The two Maxwell equations $\operatorname{div} \mathbf{B} = 0$ and $\operatorname{curl} \mathbf{E} = -\frac{\partial \mathbf{B}}{\partial t}$ are replaced, as in Chapter 9 (p. 113ff), by

$$d\mathcal{F} = 0. \tag{10-2}$$

In terms of the matrix (F_{ij}), this equation, which does not involve the metric at all, yields

$$0 = d\, F_{i<j}\, dx^i \wedge dx^j = \frac{\partial F_{i<j}}{\partial x^k}\, dx^k \wedge dx^i \wedge dx^j$$

$$= \sum_{i<j<k} \left(\frac{\partial F_{ij}}{\partial x^k} + \frac{\partial F_{ki}}{\partial x^j} + \frac{\partial F_{jk}}{\partial x^i}\right) dx^i \wedge dx^j \wedge dx^k.$$

Thus Equation 10-2 is equivalent to

$$\frac{\partial F_{ij}}{\partial x^k} + \frac{\partial F_{ki}}{\partial x^j} + \frac{\partial F_{jk}}{\partial x^i} = 0$$

$$\text{or} \tag{10-3}$$

$$F_{ij;k} + F_{ki;j} + F_{jk;i} = 0.$$

This last equation results from

$$F_{ij;k} = \frac{\partial F_{ij}}{\partial x^k} - F_{lj}\Gamma^l{}_{ki} - F_{il}\Gamma^l{}_{kj}$$

and the symmetries $\Gamma^l{}_{kj} = \Gamma^l{}_{jk}, \ldots.$

We shall assume that M^4 is time oriented, i.e., one has a continuous selection of forward light cones at each point. We shall also assume the existence of a twisted charge-current 3-form $\mathcal{S}$, dependent only on the conformal (light-cone) structure of M^4, such that if $\mathcal{U}$ is a compact portion of a space-like hypersurface,

$$Q(\mathcal{U}) = \int_{\mathcal{U}} \mathcal{S}$$

is the charge contained in $\mathcal{U}$ (using the forward light cones to orient the normal to $\mathcal{U}$).

Using the metric of M^4, there is a unique vector J such that

$$\mathscr{S} = i_J \omega^4.$$

Then the second set of Maxwell equations $\operatorname{div} \mathbf{E} = 4\pi\sigma$ and $\operatorname{curl} \mathbf{B} = 4\pi\mathbf{j} + \frac{\partial \mathbf{E}}{\partial t}$ is replaced, as in Chapter 9 (p. 109ff), by

$$d * \mathscr{F} = 4\pi \mathscr{S}, \tag{10-4}$$

which again yields conservation of charge. We can again write locally, $*\mathscr{F} = -\mathscr{H} \wedge dt + \mathscr{D}$, where $\mathscr{D}$ contains no dt terms. If we put

$$\begin{cases} \mathscr{H} = H_1\, dx^1 + H_2\, dx^2 + H_3\, dx^3 \\ \mathscr{D} = D_{23}\, dx^2 \wedge dx^3 + D_{31}\, dx^3 \wedge dx^1 + D_{12}\, dx^1 \wedge dx^2 \end{cases}$$

then, as in Equation 9-14,

$$\begin{cases} H_1 = \sqrt{|g|}\, F^{23}, H_2 = \sqrt{|g|}\, F^{31}, H_3 = \sqrt{|g|}\, F^{12} \\ D_{23} = \sqrt{|g|}\, F^{01}, D_{31} = \sqrt{|g|}\, F^{02}, D_{12} = \sqrt{|g|}\, F^{03}. \end{cases} \tag{10-5}$$

If now one writes out $d * \mathscr{F}$ in full, and compares with $4\pi\mathscr{S} = 4\pi i_J \omega^4 = 4\pi \sqrt{|g|}\, i_J\, dx^0 \wedge dx^1 \wedge dx^2 \wedge dx^3$, one sees that the four equations

$$\frac{1}{\sqrt{|g|}} \frac{\partial}{\partial x^j} (\sqrt{|g|}\, F^{ij}) = 4\pi J^i \tag{10-6}$$

are the coordinate expressions of Equation 10-4.

Equation 10-6, which is a divergence-type expression, can be written in a form involving covariant derivatives. To see this, first note that from Equation 6-8 the divergence of a vector X in a (pseudo)-Riemannian M^n is given by

$$\operatorname{div} X = \frac{1}{\sqrt{|g|}} \frac{\partial}{\partial x^i} (\sqrt{|g|}\, X^i).$$

Now at the pole of a geodesic coordinate system we have $dg_{ij} = 0$, and so at this pole $\operatorname{div} X = \partial X^i/\partial x^i$. On the other hand, the scalar $X^i_{;i} = \partial X^i/\partial x^i + X^j \Gamma^i_{ij}$ also has this same expression at the pole and so we have the well-known expression

$$\operatorname{div} X = X^i{}_{;i}. \tag{10-7}$$

Then $\dfrac{1}{\sqrt{|g|}}\dfrac{\partial}{\partial x^i}(\sqrt{|g|}\,X^i) = X^i{}_{;i}$ yields the useful relation

$$\Gamma^i{}_{ij} = \frac{\partial}{\partial x^j}\log\sqrt{|g|} \tag{10-8}$$

(this can also be seen from the well-known expression for the derivative of the determinant of a matrix).

Returning to Equation 10-6, first note that from Equation 10-8

$$\frac{1}{\sqrt{|g|}}\frac{\partial}{\partial x^j}(\sqrt{|g|}\,F^{ij}) = \frac{\partial F^{ij}}{\partial x^j} + F^{ij}\Gamma^l{}_{jl},$$

while

$$\begin{aligned} F^{ij}{}_{;j} &= \frac{\partial F^{ij}}{\partial x^j} + F^{lj}\Gamma^i{}_{jl} + F^{il}\Gamma^j{}_{jl} \\ &= \frac{\partial F^{ij}}{\partial x^j} + F^{il}\Gamma^j{}_{jl}, \end{aligned}$$

since F^{lj} is skew and $\Gamma^i{}_{jl}$ is symmetric in (l, j). Consequently,

$$\frac{1}{\sqrt{|g|}}\frac{\partial}{\partial x^j}(\sqrt{|g|}F^{ij}) = F^{ij}{}_{;j} \tag{10-9}$$

and Equation 10-6 can be written in its most common form

$$F^{ij}{}_{;j} = 4\pi J^i, \tag{10-10}$$

that is,

$$\mathrm{Div}(\mathscr{F}) = 4\pi J.$$

The Electromagnetic Stress-Energy-Momentum Tensor

The electromagnetic field influences space-time in a manner prescribed by the Einstein equations. For this we will need to know how the electromagnetic field contributes to the stress-energy-momentum tensor. Faraday associated stresses with an electromagnetic field and

Thomson and Maxwell associated an energy density to the field. Recall that if we have an energy-momentum tensor of the form $T^{ij} = \rho_0 u^i u^j + E^{ij}$, then the equations of motion yield (see Eqn. 7-24) $\rho_0 \nabla_u u = -\Pi \operatorname{Div} E = -(\operatorname{Div} E)^\perp$. Consider then the motion of a charged *dust*. Let ρ_0 be the rest density of the dust. The only stresses present will be electromagnetic. If we consider the Lorentz force formula for a test particle of charge e and rest mass m_0, $m_0(\nabla_u u)^i = eF^i{}_j u^j$, we can define the force *density* in the case of a continuous distribution of charge by dividing both sides by the rest volume:

$$\rho_0(\nabla_u u)^i = \sigma_0 F^i{}_j u^j = F^i{}_j J^j, \tag{10-11}$$

where σ_0 is the rest charge density and $J = \sigma_0 u$ is the current 4-vector. Thus we have a candidate for the stress-energy-momentum tensor of the pure electromagnetic field if we can find an (E^{ij}) such that

$$F^i{}_j J^j = -E^{ij}{}_{;j},$$

since $F^i{}_j J^j$ is automatically orthogonal to u due to the skew symmetry of (F_{ij}). Such an (E^{ij}) is easily found, as follows (Minkowski). From Equation 10-10,

$$\begin{aligned} 4\pi F^i{}_j J^j &= F^i{}_j F^{jk}{}_{;k} = (F^i{}_j F^{jk})_{;k} - F^i{}_{j;k} F^{jk} \\ &= (F^i{}_j F^{jk})_{;k} - g^{ir} F_{rj;k} F^{jk}. \end{aligned}$$

Skew symmetry of (F^{jk}) gives

$$\begin{aligned} F_{rj;k} F^{jk} &= \frac{1}{2}(F_{rj;k} - F_{rk;j}) F^{jk} \\ &= \frac{1}{2}(F_{rj;k} + F_{kr;j}) F^{jk}, \end{aligned}$$

which by Equation 10-3 yields

$$F_{rj;k} F^{jk} = -\frac{1}{2} F_{jk;r} F^{jk} = -\frac{1}{4}(F_{jk} F^{jk})_{;r}.$$

Thus

$$4\pi F^i{}_j J^j = (F^i{}_j F^{jk} + \frac{1}{4} g^{ik} F_{rs} F^{rs})_{;k}$$

and our candidate for the stress-energy-momentum tensor for the pure electromagnetic field is

$$E^{ik} = \frac{1}{4\pi}\left(F^{ij}F^{k}{}_{j} - \frac{1}{4}g^{ik}F_{rs}F^{rs}\right). \tag{10-12}$$

It is immediately apparent that the trace of this symmetric tensor vanishes, $E^{i}{}_{i} = 0$. Let $\mathsf{E} = (E^{i}{}_{j})$ be the matrix of the linear transformation corresponding to the symmetric (E^{ij}) and let $\mathsf{F} = (F^{i}{}_{j})$ that for the skew (F_{ij}). Then from Equation 10-12, $4\pi\mathsf{E} = -\mathsf{F}^2 + \frac{1}{4}\,\mathrm{tr}\,(\mathsf{F}^2)\,\mathsf{I}$.

Consider now a space like hypersurface V^3 and a point $p \in V^3$. Let $t = x^0, x^1, x^2, x^3$ be a coordinate system such that, at the point p, we have

$$e_\alpha = \partial/\partial x^\alpha, \quad \alpha = 1, 2, 3 \text{ are orthonormal,}$$

tangent to V^3, and $\partial/\partial t$ is perpendicular to V^3.

Let $e_0 = (-g_{00})^{-1/2}\,\partial/\partial t$ be the unit normal to V^3 at p. Then

$$\begin{aligned}
4\pi\,\mathsf{E}(e_0, e_0) &= 4\pi\langle \mathsf{E}e_0, e_0\rangle \\
&= \langle -\mathsf{F}^2 e_0, e_0\rangle + \frac{1}{4}\mathrm{tr}(\mathsf{F}^2)\langle e_0, e_0\rangle \\
&= \langle \mathsf{F}e_0, \mathsf{F}e_0\rangle - \frac{1}{4}\mathrm{tr}(\mathsf{F}^2) \\
&= \|\mathsf{F}e_0\|^2 - \frac{1}{4}\mathrm{tr}(\mathsf{F}^2) \\
&= \|i_{e_0}\mathcal{F}\|^2 - \frac{1}{4}\mathrm{tr}(\mathsf{F}^2) \\
&= \|i_{e_0}(\mathcal{E}\wedge dt + \mathcal{B})\|^2 - \frac{1}{4}\mathrm{tr}(\mathsf{F}^2) \\
&= \frac{1}{-g_{00}}|\mathcal{E}|^2 - \frac{1}{4}\mathrm{tr}(\mathsf{F}^2),
\end{aligned}$$

where $|\mathcal{E}|^2 = |\mathbf{E}|^2 = E_1^2 + E_2^2 + E_3^2$. Also, since $\{e_\alpha\}$ are orthonormal (i.e., $g_{\alpha\beta} = \delta_{\alpha\beta}$),

$$\begin{aligned}
-\frac{1}{4}\mathrm{tr}(\mathsf{F}^2) &= \frac{1}{2}F_{r<s}F^{rs} \\
&= \frac{1}{2}(F_{0\alpha}F^{0\alpha} + F_{\alpha<\beta}F^{\alpha\beta})
\end{aligned}$$

$$= \frac{1}{2}\,(g^{00}F_{0\alpha}F_{0\alpha} + F_{\alpha<\beta}F^{\alpha\beta})$$

$$= \frac{1}{2}\,(g^{00}|\mathbf{E}|^2 + |\mathbf{B}|^2),$$

hence

$$4\pi \mathsf{E}(e_0, e_0) = -g^{00}|\mathbf{E}|^2 + \frac{1}{2}\,(g^{00}|\mathbf{E}|^2 + |\mathbf{B}|^2),$$

that is,

$$\mathsf{E}(e_0, e_0) = \frac{-g^{00}|\mathbf{E}|^2 + |\mathbf{B}|^2}{8\pi}. \tag{10-13}$$

In Minkowski space this reduces to the classical expression for the *energy density* of the pure electromagnetic field

$$\frac{1}{8\pi}\,(|\mathbf{E}|^2 + |\mathbf{B}|^2).$$

A pure electromagnetic field curves space-time. In the special case of a static universe corresponding to a time-independent electromagnetic field, the scalar curvature of the spatial section is, as given by Equation 4-16,

$$R_{V^3} = 16\pi\kappa\, \mathsf{E}(e_0, e_0) = 2\kappa(-g^{00}|\mathbf{E}|^2 + |\mathbf{B}|^2).$$

The Reissner Solution

The Maxwell equation $d\mathscr{F} = 0$ implies, via the converse of the Poincaré lemma, the local existence of a 1-form $\mathscr{A}$ such that $\mathscr{F} = d\mathscr{A}$. The 1-form $\mathscr{A}$ is not unique, allowing "gauge transformations" of the form $\mathscr{A} \to \mathscr{A} + df$, f any function. In terms of coordinates, a given $\mathscr{A}$ splits uniquely as $\mathscr{A} = \varphi\, dt + \boldsymbol{\mathscr{A}}$, where $\boldsymbol{\mathscr{A}} = A_\alpha\, dx^\alpha$ and (using the notation of Chapter 9, p. 109ff)

$$d\mathscr{A} = \mathbf{d}\varphi \wedge dt + \mathbf{d}\boldsymbol{\mathscr{A}} + dt \wedge \partial \boldsymbol{\mathscr{A}}/\partial t.$$

Comparison with $\mathscr{F} = \mathscr{E} \wedge dt + \mathscr{B}$ shows

$$\begin{cases} \mathscr{E} = \mathbf{d}\varphi - \dfrac{\partial \mathscr{A}}{\partial t} \\ \mathscr{B} = \mathbf{d}\mathscr{A}, \end{cases} \tag{10-14}$$

where φ is the *scalar* potential and $\mathscr{A}$ the *vector* potential.

Consider now the special case of a static universe

$$\begin{cases} ds^2 = g_{00}\, dt^2 + dl^2 \\ dl^2 = g_{\alpha\beta}\, dx^\alpha\, dx^\beta; \text{ spatial metric on } V^3 \end{cases}$$

and a time-independent electromagnetic field. Then $\mathscr{E} = \mathbf{d}\varphi = d\varphi$ and $\mathscr{B} = d\mathscr{A}$. Maxwell's equations (Eqn. 10-6) yield, for $i = 0$,

$$4\pi J^0 = \frac{1}{\sqrt{|g|}} \frac{\partial}{\partial x^j} (\sqrt{|g|}\, F^{0j}).$$

Putting $J = \sigma_0 u$, i.e., $J^0 = \sigma_0 u^0 = \sigma_0/\sqrt{-g_{00}}$, and $\sqrt{|g|} = \sqrt{-g_{00}}\, \sqrt{g_V}$, and using the fact that everything is time independent, we have

$$\begin{aligned} 4\pi\sigma_0 &= \frac{1}{\sqrt{g_V}} \frac{\partial}{\partial x^\alpha} \left(\frac{1}{\sqrt{-g_{00}}} \sqrt{g_V}\, g^{\alpha\beta} \frac{\partial \varphi}{\partial x^\beta} \right) \\ &= \operatorname{div}_V \left(\frac{1}{\sqrt{-g_{00}}} \mathbf{E} \right) = \operatorname{div}_V \left(\frac{1}{\sqrt{-g_{00}}} \operatorname{grad}_V \varphi \right) \end{aligned} \tag{10-15}$$

where, as usual, $E^\alpha = g^{\alpha\beta} E_\beta = g^{\alpha\beta}\, \partial\varphi/\partial x^\beta$. We shall return later for a discussion of Equation 10-15 and its relation to the classical Poisson equation.

Consider now, with Reissner (1916), the special case of a field due to a single particle of mass m and charge e. As in the Schwarzschild case we consider the particle at the origin and invoke spherical symmetry $ds^2 = g_{00}(r)\, dt^2 + g_{rr}(r)\, dr^2 + r^2\, d\Omega^2$. Outside the particle (whose coordinate radius will be written r_0) the stress-energy-momentum tensor is that of a pure electrostatic field, and we can assume, by symmetry, that this field is of the form $\mathscr{F} = \mathscr{E} \wedge dt = \partial\varphi/\partial r\, dr \wedge dt$, $\varphi = \varphi(r)$, i.e., $F_{r0} = \partial\varphi/\partial r = -F_{0r}$, and all other components vanish. Automatically $d\mathscr{F} = 0$, so that the first set of Maxwell equations is satisfied outside the particle.

To satisfy the second set of Maxwell equations we consider a ball B_R centered at the origin, with coordinate radius $R > r_0$. From Equation 10-15, spherical symmetry, and Equation 6-11 we have

$$4\pi e = \int_{B_R} \operatorname{div}_V \left(\frac{1}{\sqrt{-g_{00}}} \mathbf{E} \right) \omega_V = \int_{\partial B_R} \frac{1}{\sqrt{-g_{00}}} i_{\mathbf{E}} \omega_V$$

$$= 4\pi R^2 \frac{|\mathbf{E}(R)|}{\sqrt{-g_{00}(R)}}.$$

Consequently

$$|\mathbf{E}(r)| = \sqrt{-g_{00}(r)} \frac{e}{r^2},$$

and since $\partial/\partial r$ has length $\sqrt{g_{rr}}$

$$E^r = \frac{\sqrt{-g_{00}(r)}}{\sqrt{g_{rr}(r)}} \cdot \frac{e}{r^2}.$$

Since $J = \sigma_0 u$ vanishes outside the particle and since the electrical and metrical fields are time independent, it is easy to see that the only non-vacuous equation in the second set of Maxwell's equations (Eqn. 10-6) is Equation 10-15, for $r > r_0$. One verifies immediately that the above expression for E^r satisfies this equation.

For the stress-energy-momentum tensor outside the particle we have, from Equation 10-13,

$$\mathsf{E}(u, u) = -\frac{1}{8\pi} g^{00} |\mathbf{E}|^2 = \frac{1}{8\pi} \frac{e^2}{r^4}, \tag{10-16}$$

which is the same as the classical expression for the energy density of this electrostatic field.

For $r > r_0$ we may compute g_{rr} by means of Equation 5-7, with ρ replaced by $\mathsf{E}(u, u)$. If the constant α is again interpreted as the mass of the particle, as in the Schwarzschild solution, we obtain the *spatial* Reissner metric

$$\begin{cases} dl^2 = g_{rr}\, dr^2 + r^2(d\theta^2 + \sin^2\theta\, d\varphi^2) \\ g_{rr} = \left(1 - \dfrac{2m}{r} + \dfrac{\kappa e^2}{r^2}\right)^{-1}. \end{cases} \tag{10-17}$$

We next need to compute g_{00} for the region $r > r_0$ outside the particle. Let $e_0 = u = (-g_{00})^{-1/2}\, \partial/\partial t$, and let e_r, e_θ, e_φ be the spatial unit orthogonal vectors adapted to the spherical coordinate system. As we have seen, $\mathsf{E}(e_0, e_0) = \rho = (8\pi)^{-1} e^2 r^{-4}$. The same type of calculation shows

$$\mathsf{E}(e_r, e_r) = -\rho, \qquad \mathsf{E}(e_\theta, e_\theta) = \mathsf{E}(e_\varphi, e_\varphi) = \rho,$$

the negatives of the classical *Faraday stresses*. We now use the second Einstein equation

$$\mathsf{K}(e_\theta \wedge e_\varphi) + \mathsf{K}(e_0 \wedge e_\theta) + \mathsf{K}(e_0 \wedge e_\varphi) = -8\pi\kappa \mathsf{E}(e_r, e_r).$$

As in the Schwarzschild solution, we invoke Equation 11-10 (with pressure p_r replaced by Faraday stress $-\rho$)

$$\frac{-\kappa e^2}{r^4} = 8\pi\kappa \mathsf{E}(e_r, e_r) = -\frac{1}{r^2} + \frac{1}{r^2 g_{rr}} + \frac{2}{r g_{rr}} \frac{\partial \log \sqrt{-g_{00}}}{\partial r}.$$

Replacing g_{rr} by its Reissner value (Eqn. 10-17) yields a differential equation that is easily solved to give

$$-g_{00} = \frac{1}{g_{rr}} = 1 - \frac{2m}{r} + \frac{\kappa e^2}{r^2}.$$

This completes the Reissner solution,

$$ds^2 = -\left(1 - \frac{2m}{r} + \frac{\kappa e^2}{r^2}\right) dt^2 + \frac{dr^2}{1 - \dfrac{2m}{r} + \dfrac{\kappa e^2}{r^2}} + r^2\, d\Omega^2. \qquad (10\text{-}18)$$

(For a discussion of the "singularity" where $g_{00} = 0$, see Hawking and Ellis, 1973.)

Conformal Invariance of Maxwell's Equations

In the space-time of general relativity we have postulated that an electromagnetic field gives rise to a 2-form $\mathcal{F}$ satisfying

$$\begin{cases} d\mathcal{F} = 0 \\ d{*}\mathcal{F} = 4\pi\mathcal{S}. \end{cases}$$

Suppose now that one keeps $\mathcal{F}$ fixed but allows the metric (g_{ij}) of space-time to change conformally. The first equation $d\mathcal{F} = 0$ still holds, since the operator d is independent of metric. Also, the form $\mathcal{S}$ was postulated to depend only on the conformal structure (though the current vector J

depends on the volume element). Note (from Eqn. 9-9) that the Hodge $*$-operator depends on the metric used, and so the equation $d{*}\mathscr{F} = 4\pi\mathscr{S}$ must be investigated. However, if we have two conformally related metrics in a (pseudo)-Riemannian M^{2n}, $d\tilde{s}^2 = \lambda^2\, ds^2$, then one sees easily from Equation 9-9 that while $\tilde{*} \neq *$ in general, it is true that when operating on forms of degree n (the *middle dimension*), $\tilde{*} = *$. Consequently, if the two metrics of M^4 are conformally related, $\tilde{*}\mathscr{F} = *\mathscr{F}$, and so *Maxwell's equations are conformally invariant!* (For Minkowski space this is due to Cunningham and to Bateman in 1910.)

Poisson's Equation in a Static Universe

The equations in 10-14 are classically written in 3-space as

$$\begin{cases} \mathbf{E} = \text{grad}\ \varphi - \dfrac{\partial \mathbf{A}}{\partial t} \\ \\ \mathbf{B} = \text{curl}\ \mathbf{A}, \end{cases}$$

while Gauss's law is

$$\text{div}\ \mathbf{E} = 4\pi\sigma,$$

and so

$$4\pi\sigma = \text{div grad}\ \varphi - \frac{\partial}{\partial t}\,\text{div}\, A\,.$$

In particular, for a time-independent electromagnetic field, φ satisfies Poisson's equation $\nabla^2\varphi = 4\pi\sigma$. In general relativity, for a time-independent electromagnetic field, we still have $\mathbf{E} = \text{grad}_V\varphi$, but Gauss's law is replaced by Equation 10-15, $4\pi\sigma = \text{div}_V\left(\dfrac{1}{\sqrt{-g_{00}}}\ \text{grad}_V\varphi\right)$, and the right side is no longer a spatial Laplacian because of the factor $\dfrac{1}{\sqrt{-g_{00}}}\cdot$ Thus φ does *not* satisfy Poisson's equation.

In Chapter 8 we saw the usefulness of the Fermat metric $dl_F = \dfrac{1}{\sqrt{-g_{00}}}\,dl$ in considering the behavior of light rays in the spatial sections of a static universe. Since light is an electromagnetic phenomenon it is reasonable to believe that the Fermat metric might be useful when considering time-independent electromagnetic fields. If we introduce the Fermat metric

$$g^F_{\alpha\beta} = \frac{g_{\alpha\beta}}{-g_{00}}, \; g_F^{\alpha\beta} = -g_{00}g^{\alpha\beta},$$

then, since charge is independent of metric, the Fermat charge density σ_F must satisfy $\sigma_F \sqrt{g_F} = \sigma \sqrt{g_V}$, that is,

$$\sigma_F = \sigma(-g_{00})^{3/2}.$$

One sees then immediately from Equation 10-15 that

$$4\pi\sigma_F = \frac{1}{\sqrt{g_F}} \frac{\partial}{\partial x^\alpha} \left(\sqrt{g_F}\, g_F^{\alpha\beta} \frac{\partial \varphi}{\partial x^\beta} \right) = \nabla_F^2 \varphi,$$

where ∇_F^2 is the Laplacian in the Fermat metric. Thus φ *satisfies Poisson's equation in the Fermat metric!*

J. Ehlers pointed out that the essential ingredient for this fact is the conformal invariance of Maxwell's equations. Indeed, if one introduces the Fermat metric into the full space-time manifold, locally

$$ds_F^2 = \frac{ds^2}{-g_{00}} = -dt^2 + dl_F^2,$$

we may apply Equation 10-15 directly to this metric, replacing σ by σ_F. Since in the new metric $g^F_{00} \equiv -1$, we recover $\nabla_F^2 \varphi = 4\pi\sigma_F$.

As an immediate corollary, we conclude that in the absence of charges in this static case the electrostatic potential φ is *Fermat harmonic* (the reader may check this directly in the case of the Reissner solution). In particular, we recover the (nonrelativistic) result that the electrostatic potential in a charge-free, compact, connected cavity inside a conductor is constant (since $\mathbf{E} = 0$ in a conductor and φ satisfies the maximum principle in the cavity).

The physical significance of the Fermat metric seems elusive, but mathematically it does shed some light. Perhaps this indicates that whenever a nongravitational field is introduced into general relativity, it might help to introduce a suitable (conformally?) related metric into the spatial sections, at least in the static case.

Nonstatic Fields

I shall conclude with a very brief note on potentials for time-dependent electromagnetic fields. Here it is important to use the full four-dimensional differential operators. Introduce the Hodge operator $\delta = -*d*$, which is the negative of the four-dimensional divergence operator for forms, and the "Laplace operator" (introduced by Weitzenböck) $\Delta = \delta d + d\delta$, which has the opposite sign to the usual Laplace operator of vector analysis in the *Riemannian* case. From $4\pi\mathcal{S} = d * \mathcal{F} = d * (d\mathcal{A})$ we get $4\pi * \mathcal{S} = -\delta\, d\mathcal{A} = -\Delta\mathcal{A} + d\delta\mathcal{A}$. It can be shown that locally one can choose the "Lorentz gauge," i.e., $\delta\mathcal{A} = 0$, and we then have that $\mathcal{A}$ satisfies the "wave equation"

$$\Box^2\mathcal{A} = 4\pi * \mathcal{S} = 4\pi * (i_J\omega^4), \tag{10-19}$$

where $\Box^2 = -\Delta$ is the generalized d'Alembertian. From Equation 9-11 and the fact that $** = -1$ in our case,

$$*(i_J\omega^4) = **(J_k\, dx^k) = -J_k\, dx^k.$$

If we use the Weitzenböck coordinate expression for Δ (see de Rham, 1955, but note that de Rham's Riemann tensor has the opposite sign to ours), then Equation 10-19 can be written

$$g^{ij}\mathcal{A}_{k;ij} - R^h{}_k\mathcal{A}_h = -4\pi J_k. \tag{10-20}$$

II

The Interior Solution

Curvature of World Lines and Gravitational Force Potential

In this chapter we shall investigate in more detail the geometry of space-time in regions occupied by matter or energy, a study of great importance in understanding the relativistic effects involved in stellar structure. Although we shall be concerned almost entirely with static situations, we shall begin by considering general metrics where the time lines are orthogonal to the spatial sections V_t^3. This is merely a condition on local coordinates; it can always be achieved by introducing Gaussian coordinate systems (see p. 44):

$$\begin{cases} ds^2 = g_{00}(t, x)\, dt^2 + dl^2 \\ dl^2 = g_{\alpha\beta}(t, x)\, dx^\alpha\, dx^\beta \text{ is the metric on } V_t^3. \end{cases}$$

Consider the geodesic curvature of the t-lines, $\nabla_{e_0}e_0$, where $e_0 = (-g_{00})^{-1/2}\partial/\partial t$ is the unit tangent field to the t-lines. Recalling that the geodesic curvature of a world line corresponds to the acceleration vector in mechanics (p. 83ff), and in Newtonian theory the acceleration is the spatial gradient of the Newtonian gravitational potential, the following lemma states that $-\log\sqrt{-g_{00}}$ is the analog of the Newtonian potential as far as *force* is concerned. (In Chapter 2 we indicated that $1-\sqrt{-g_{00}}$ is the analog of the Newtonian potential as far as *potential energy* is concerned.)

Lemma The geodesic curvature of a t-line is

$$\nabla_{e_0}e_0 = \mathrm{grad}_V \log\sqrt{-g_{00}}.$$

PROOF Let $P = (t, x)$ be a given event on V_t^3. Let e_1, e_2, e_3 be tangent fields to V_t^3 that are orthonormal at P. Extend e_1, e_2, e_3 along the t-lines through V_t^3 by making them invariant under time translation (see p. 57ff), that is,

$$[\partial/\partial t, e_\alpha] = \nabla_{\partial/\partial t}e_\alpha - \nabla_{e_\alpha}\partial/\partial t = 0.$$

Then, at P, we have $\langle\nabla_{e_0}e_0, e_\alpha\rangle = e_0\langle e_0, e_\alpha\rangle - \langle e_0, \nabla_{e_0}e_\alpha\rangle = -\langle e_0, \nabla_{e_0}e_\alpha\rangle$ since e_α remains tangent to each V_t^3. Then, setting $\varphi = \sqrt{-g_{00}}$, we have

$$\begin{aligned}\langle\nabla_{e_0}e_0, e_\alpha\rangle &= -\varphi^{-1}\langle e_0, \nabla_{\partial/\partial t}e_\alpha\rangle\\ &= -\varphi^{-1}\langle e_0, \nabla_{e_\alpha}\partial/\partial t\rangle\\ &= \langle e_0, e_\alpha(\varphi^{-1})\partial/\partial t\rangle\\ &= -\varphi^{-2}e_\alpha(\varphi)\langle e_0, \partial/\partial t\rangle\\ &= \varphi^{-1}e_\alpha(\varphi).\end{aligned}$$

Since $\nabla_{e_0}e_0$ is orthogonal to e_0, we have, at P

$$\begin{aligned}\nabla_{e_0}e_0 &= \sum_{\alpha=1}^{3}\langle\nabla_{e_0}e_0, e_\alpha\rangle e_\alpha = \sum_{\alpha=1}^{3}\varphi^{-1}e_\alpha(\varphi)e_\alpha\\ &= \sum_{\alpha=1}^{3}e_\alpha(\log\sqrt{-g_{00}})e_\alpha \equiv \mathrm{grad}_V \log\sqrt{-g_{00}},\end{aligned}$$

as desired.

The Schwarzschild Interior Solution

Consider a perfect fluid at rest in a *static* universe. The unit velocity vector u of the fluid is orthogonal to the spatial sections $V_t^3 = V^3$. From the above lemma and the equations of motion (Eqn. 7-26)

$$(\rho + p)\mathsf{grad}_V \log \sqrt{-g_{00}} = -\mathsf{grad}_V p, \qquad (11\text{-}1)$$

which is an equation of hydrostatic equilibrium relating the pressure gradient to the gradient of a type of gravitational potential, $\log \sqrt{-g_{00}}$.

Suppose now that the static universe is spherically symmetric (spatially) about a point O. As in Chapter 5 we introduce spherical coordinates with origin at O

$$ds^2 = g_{00}(r)\, dt^2 + g_{rr}(r)\, dr^2 + r^2(d\theta^2 + \sin^2\theta\, d\varphi^2),$$

and again we have

$$g_{rr} = \frac{1}{1 - \dfrac{2m(r)}{r}},$$

where

$$\begin{cases} m(r) = \kappa \displaystyle\int_0^r 4\pi(r')^2\rho(r')\, dr' \\ \rho(r) = \mathsf{T}(u, u). \end{cases}$$

The spatial metric is completely determined by the density ρ alone, but the g_{00} component has been determined, in Chapter 5, only in the region outside of matter (the "exterior" region); if the energy-momentum-tensor vanishes for $r > r_0$, then

$$-g_{00}(r) = 1 - \frac{2m(r_0)}{r}, \qquad r > r_0.$$

Consider now (with Schwarzschild) the case of a fluid ball of coordinate "radius" r_0 with *constant density* ρ. From Equation 4-16 we know $R_V = 16\pi\kappa\rho$ is constant, for $r < r_0$; but actually we can say much more, for in the case of constant ρ

$$g_{rr}(r) = \frac{1}{1 - \dfrac{8\pi\kappa\rho}{3} r^2} = \frac{1}{1 - \dfrac{r^2}{\tilde{R}^2}},$$

where

$$\tilde{R}^2 \equiv \frac{3}{8\pi\kappa\rho} = \frac{6}{R_V}. \tag{11-2}$$

But the metric

$$dl^2 = \frac{dr^2}{1 - \dfrac{r^2}{\tilde{R}^2}} + r^2(d\theta^2 + \sin^2\theta\, d\varphi^2)$$

is easily seen to be the standard metric on the 3-sphere of radius $\tilde{R}$ in Euclidean 4-space, whose metric is $\tilde{R}^2\, d\psi^2 + r^2(d\theta^2 + \sin^2\theta\, d\varphi^2)$. Here $r = \tilde{R} \sin \psi$. In Figure 11-1 we have joined this interior, 3-spherical cap solution to the exterior Schwarzschild solution, a Flamm paraboloid. At the join we have

$$\frac{r_0^2}{\tilde{R}^2} = \frac{2m}{r_0} \tag{11-3}$$

to match the two expressions for $g_{rr}(r_0)$. We also note that the equality given in Equation 11-2 (i.e., $R_V = 6/\tilde{R}^2$) is simply the expression of the fact that the 3-sphere of radius $\tilde{R}$ has constant sectional curvature

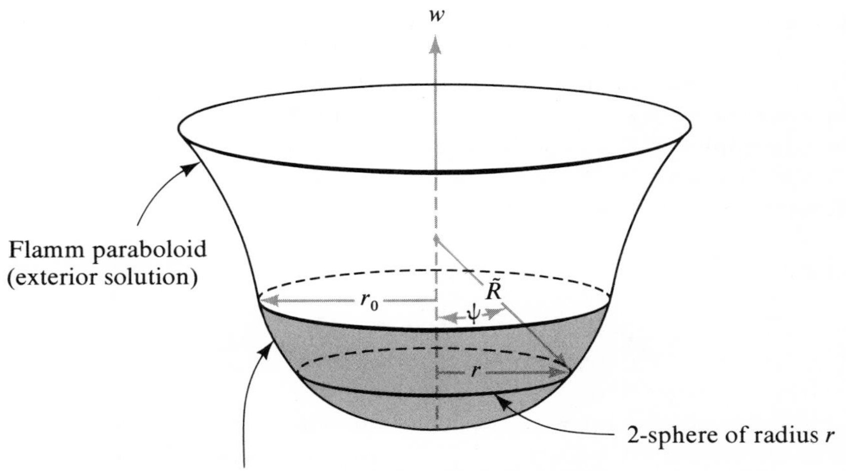

FIGURE 11-1

$K = 1/\tilde{R}^2$, that the Ricci curvature is a sum of two sectional curvatures, and that the scalar curvature R_V is a sum of three Ricci curvatures.

We must now determine the interior version of g_{00} that must match up with $-(1 - 2m/r)$ at $r = r_0$. For notational convenience let us put $\xi(r) = \sqrt{-g_{00}(r)}$. From Equation 11-1, $(\rho + p)d/dr \log \xi = -dp/dr$, since p and ξ are functions of r alone, by spherical symmetry. Since ρ is constant, this is immediately solved to give $p = \text{constant } \xi^{-1} - \rho$. Since the pressure is assumed to vanish at the surface of the liquid, $r = r_0$, we get, for $r \leq r_0$

$$p(r) = \frac{\rho\xi(r_0)}{\xi(r)} - \rho = \rho\left(\frac{\sqrt{-g_{00}(r_0)}}{\sqrt{-g_{00}(r)}} - 1\right), \tag{11-4}$$

relating the pressure to the potential.

Now we consider the generalized Poisson equation (Eqn. 3-9) and use the above expression for p:

$$\begin{aligned}\nabla_V^2 \sqrt{-g_{00}} = \nabla_V^2\xi &= 4\pi\kappa(\rho + 3\mathrm{p})\xi \\ &= -8\pi\kappa\rho\xi + 12\pi\kappa\rho\xi(r_0),\end{aligned}$$

that is, $\xi(r) = \sqrt{-g_{00}(r)}$ must satisfy the equation

$$\nabla_V^2\xi = -b\xi + a, \tag{11-5}$$

where $a = 12\pi\kappa\rho\xi(r_0)$, $b = 8\pi\kappa\rho$. A particular solution of Equation 11-5 is clearly $\xi = a/b$. We need only consider then the homogeneous equation $\nabla_V^2\tilde{\xi} = -b\tilde{\xi}$, showing that $\tilde{\xi}$ must be an eigenfunction of the Laplacian on the 3-sphere, with eigenvalue $-8\pi\kappa\rho$. For functions of $r = \tilde{R}\sin\psi$ alone, this eigenfunction equation becomes, in the metric

$$dl^2 = \tilde{R}^2\, d\psi^2 + \tilde{R}^2 \sin^2\psi(d\theta^2 + \sin^2\theta\, d\varphi^2),$$

the equation

$$\nabla_V^2\tilde{\xi} = \frac{1}{\tilde{R}^2\sin^2\psi}\frac{\partial}{\partial\psi}\left(\sin^2\psi\,\frac{\partial\tilde{\xi}}{\partial\psi}\right) = -b\tilde{\xi},$$

with immediate solution $\tilde{\xi}(\psi) = \text{constant} \cdot \cos\psi$, using the fact that $3/\tilde{R}^2 = 8\pi\kappa\rho = b$ by Equation 11-2. The general solution of Equation 11-5 then is (A a constant)

$$\xi(\psi) = A \cdot \cos\psi + a/b = A\cos\psi + \frac{3}{2}\,\xi(\psi_0),$$

where $r_0 = \tilde{R} \sin \psi_0$. From Equation 11-3

$$A \cos \psi_0 = -\frac{1}{2}\,\xi(\psi_0) = -\frac{1}{2}\sqrt{1 - \frac{2m}{r_0}}$$
$$= -\frac{1}{2}\sqrt{1 - \frac{r_0^2}{\tilde{R}^2}} = -\frac{1}{2}\cos\psi_0,$$

and so $A = -\frac{1}{2}$. Consequently,

$$\begin{aligned}\sqrt{-g_{00}(\psi)} &= -\frac{1}{2}\cos\psi + \frac{3}{2}\cos\psi_0\\ &= -\frac{1}{2}\sqrt{1 - \frac{r^2}{\tilde{R}^2}} + \frac{3}{2}\sqrt{1 - \frac{r_0^2}{\tilde{R}^2}}\end{aligned} \tag{11-6}$$

and we have the *interior Schwarzschild solution,* for $r < r_0$

$$\begin{aligned}ds^2 &= -\left(-\frac{1}{2}\cos\psi + \frac{3}{2}\cos\psi_0\right)^2 dt^2 + \tilde{R}^2\, d\psi^2\\ &\quad + \tilde{R}^2 \sin^2\psi(d\theta^2 + \sin^2\theta\, d\varphi^2)\\ &= -\left(-\frac{1}{2}\sqrt{1 - \frac{r^2}{\tilde{R}^2}} + \frac{3}{2}\sqrt{1 - \frac{r_0^2}{\tilde{R}^2}}\right)^2 dt^2 + \frac{dr^2}{1 - \frac{r^2}{\tilde{R}^2}}\\ &\quad + r^2(d\theta^2 + \sin^2\theta\, d\varphi^2).\end{aligned} \tag{11-7}$$

This expression only makes sense as long as $-g_{00}(\psi) = (-\tfrac{1}{2}\cos\psi + \tfrac{3}{2}\cos\psi_0)^2$ is nonvanishing. It might be that its vanishing corresponds only to a coordinate singularity, as in the Schwarzschild "singularity." We can see, however, from Equation 11-4, that the pressure becomes infinite at a zero of g_{00}, giving indeed a physical singularity. To avoid this we must assume that g_{00} never vanishes in the interior. From Equation 11-6 we see that, in order to avoid $g_{00}(\psi)$ vanishing at $\psi = 0$, we must have $\cos\psi_0 > \tfrac{1}{3}$, i.e., from Equation 11-3,

$$\frac{2m}{r_0} = \frac{r_0^2}{\tilde{R}^2} = \sin^2\psi_0 < \frac{8}{9}.$$

(For further discussions see, for example, Misner, Thorne, Wheeler, 1973, Box 23.2.)

The Oppenheimer–Volkoff Equation

We now leave this example of Schwarzschild and proceed to a general discussion of the interior solution of a *static, spherically symmetric* mass-energy distribution. The metric is in the usual form

$$ds^2 = g_{00}(r)\,dt^2 + g_{rr}(r)\,dr^2 + r^2(d\theta^2 + \sin^2\theta\,d\varphi^2).$$

We have an orthonormal frame

$$e_0 = \frac{\partial/\partial t}{\sqrt{-g_{00}}}, \qquad e_1 = \frac{\partial/\partial r}{\sqrt{g_{rr}}} = e_r,$$

$$e_2 = \frac{\partial/\partial\theta}{r} = e_\theta, \quad e_3 = \frac{\partial/\partial\varphi}{r\sin\theta} = e_\varphi.$$

It is easy to see that these vectors are eigenvectors of the stress-energy-momentum tensor, i.e., the Einstein tensor, as follows. Let $\mathsf{G} = G^j{}_i$ be the linear transformation defined by the "mixed" Einstein tensor. Since G is constructed from the metric it is invariant under isometries in the sense that if τ is an isometry of space-time and τ_* the induced differential action on tangent vectors, then $\mathsf{G}\tau_* = \tau_*\mathsf{G}$. Let $\tau: M^4 \to M^4$ be the isometry $\tau(t, x) = (-t, x)$, which has the spatial section V_0 as fixed point set. If v is tangent to V_0, then $\tau_*\mathsf{G}v = \mathsf{G}\tau_* v = \mathsf{G}v$ shows that $\mathsf{G}v$ is also tangent to V_0, i.e., G leaves the tangent spaces to V_0 invariant. Since each such tangent space is a positive definite inner product space and G is self-adjoint, we conclude that G has three mutually orthogonal eigenvectors in each of these tangent spaces. Spherical symmetry (i.e., using other isometries) shows that e_r, e_θ, and e_φ are eigenvectors. We let $p_1 = p_r$, $p_2 = p_\theta$, $p_3 = p_\varphi$ be the corresponding eigenvalues of the stress-energy-momentum tensor. Spherical symmetry shows that $p_\theta = p_\varphi$, and these eigenvalues

$$\mathsf{T}(e_r, e_r) = p_r, \qquad \mathsf{T}(e_\theta, e_\theta) = p_\theta = p_\varphi = \mathsf{T}(e_\varphi, e_\varphi)$$

are called generalized *pressures*. Finally, e_0 must again be an eigenvector; the negative of its eigenvalue is again called the *energy density* $\rho = \mathsf{T}(e_0, e_0)$.

We now study the Einstein equation associated with $\mathsf{T}(e_r, e_r) = p_r$. Equation 4-13 gives us the following prescription. Let W_r^3 be the three-dimensional submanifold of space-time M^4 defined by putting r = constant $\neq 0$; its unit normal is e_r. We shall see in a moment that e_0, e_θ, and

e_φ are principal directions on W_r^3. If κ_0, κ_θ, κ_φ are the principal curvatures, then the Einstein equations give

$$8\pi\kappa \mathsf{T}(e_r, e_r) = 8\pi\kappa p_r = -\frac{1}{2} R_W + \kappa_0\kappa_\theta + \kappa_0\kappa_\varphi + \kappa_\theta\kappa_\varphi. \qquad (11\text{-}8)$$

Let b_W be the second fundamental form for W_r^3; thus if X is a tangent vector to W, $\mathsf{b}_W(X) = -\nabla_X e_r$. The principal curvatures of W are the eigenvalues of b_W (Figure 11-2).

First note that since each V_t^3 is totally geodesic in M^4 its unit normal e_0 must be parallel displaced along V_t^3; thus $\nabla_{e_r} e_0 = 0$ and consequently (since nothing depends on t)

$$\nabla_{e_0} e_r = \nabla_{e_r} e_0 + [e_0, e_r] = \left[\frac{\partial/\partial t}{\sqrt{-g_{00}}}, \frac{\partial/\partial r}{\sqrt{g_{rr}}}\right]$$

$$= -\frac{1}{\sqrt{g_{rr}}}\frac{\partial}{\partial r}\left(\frac{1}{\sqrt{-g_{00}}}\right)\frac{\partial}{\partial t} = \frac{1}{\sqrt{g_{rr}}}\frac{\partial}{\partial r}(\log\sqrt{-g_{00}})e_0$$

and so $\mathsf{b}_W(e_0) = -\dfrac{1}{\sqrt{g_{rr}}}\dfrac{d}{dr}(\log\sqrt{-g_{00}})e_0$ shows that e_0 is indeed an eigenvector of b_W with eigenvalue

$$\kappa_0 = -\frac{1}{\sqrt{g_{rr}}}\frac{d}{dr}(\log\sqrt{-g_{00}}).$$

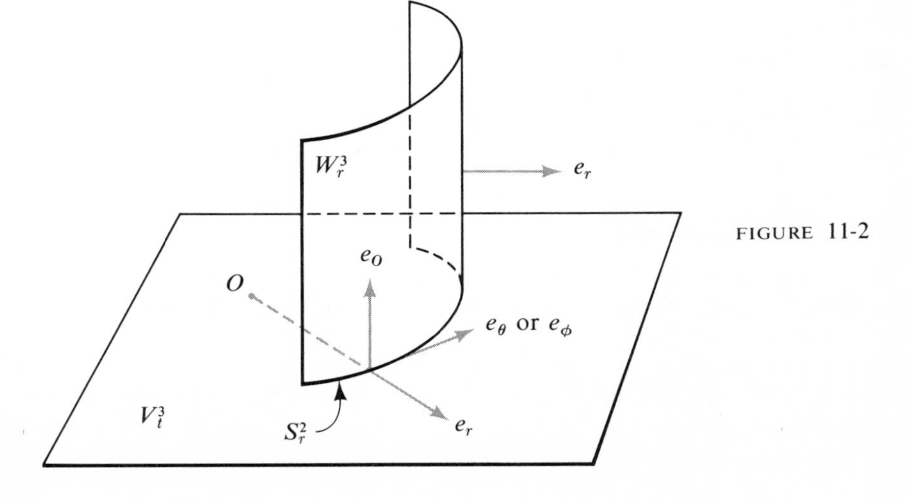

FIGURE 11-2

Note that in our spherically symmetric case the lemma on page 129 yields

$$\nabla_{e_0} e_0 = \mathsf{grad}_V \log\sqrt{-g_{00}} = g^{rr} \frac{d}{dr} \log\sqrt{-g_{00}} \frac{\partial}{\partial r}$$

$$= \frac{1}{\sqrt{g_{rr}}} \frac{d}{dr} (\log\sqrt{-g_{00}}) \cdot e_r$$

i.e., κ_0 is, in magnitude, equal to the magnitude of the geodesic curvature of the t-line, as it must.

Since b_W is self-adjoint, the remaining two eigenvectors are orthogonal to e_0, i.e., tangent to $W_r \cap V_t$, which is a 2-sphere S_r^2. By symmetry these eigenvectors can be chosen to be e_θ and e_φ, with equal eigenvalues. Since the metric is independent of φ, the "2-plane" φ = constant is a totally geodesic surface in the totally geodesic V_t^3. Since the radial lines (with tangent e_r) are geodesics in this "2-plane," and since e_r and e_θ are orthonormal, we conclude e_θ is parallel displaced along the radial lines, and so $\nabla_{e_r} e_\theta = 0$. Thus

$$\mathsf{b}_W(e_\theta) = -\nabla_{e_\theta} e_r = -\nabla_{e_r} e_\theta - [e_\theta, e_r] = [e_r, e_\theta]$$

$$= \left[\frac{\partial/\partial r}{\sqrt{g_{rr}}}, \frac{\partial/\partial\theta}{r}\right] = \frac{1}{\sqrt{g_{rr}}} \frac{d}{dr} \left(\frac{1}{r}\right) \frac{\partial}{\partial\theta} = -\frac{1}{r\sqrt{g_{rr}}} e_\theta,$$

and so

$$\kappa_\theta = \kappa_\varphi = -\frac{1}{r\sqrt{g_{rr}}}.$$

Thus

$$\kappa_0\kappa_\theta + \kappa_0\kappa_\varphi + \kappa_\theta\kappa_\varphi = \frac{2}{rg_{rr}} \frac{d}{dr} (\log\sqrt{-g_{00}}) + \frac{1}{r^2 g_{rr}}. \tag{11-9}$$

Finally, we need to compute the scalar curvature

$$\frac{1}{2} R_{W^3} = \mathsf{K}_W(e_0 \wedge e_\theta) + \mathsf{K}_W(e_0 \wedge e_\varphi) + \mathsf{K}_W(e_\theta \wedge e_\varphi)$$

for the metric induced on W_r^3

$$dl_r^2 = g_{00}(r)\, dt^2 + r^2(d\theta^2 + \sin^2\theta\, d\varphi^2),$$

where r is constant. Consequently, W_r^3 carries a *product metric* for $R \times S_r^2$. Thus the cylinder r = constant, $\theta = \pi/2$ is a totally geodesic surface in W_r^3 because its projection on S_r^2 is the (geodesic) equator of S_r^2. The metric in this cylinder is $g_{00}(r)\, dt^2 + r^2\, d\varphi^2$, which is *flat* since r is constant. Thus

$$\mathsf{K}_W(e_0 \wedge e_\varphi) = 0,$$

and by symmetry

$$\mathsf{K}_W(e_0 \wedge e_\theta) = 0.$$

Also, S_r^2 is a totally geodesic surface in the product $W_r^3 = R \times S_r^2$, and S_r^2 carries the metric of the standard 2-sphere of radius r in R^3, hence

$$\mathsf{K}_W(e_\theta \wedge e_\varphi) = \frac{1}{r^2}.$$

Consequently the Einstein equation (Eqn. 11-8) becomes

$$8\pi\kappa p_r = -\frac{1}{r^2} + \frac{1}{r^2 g_{rr}} + \frac{2}{r g_{rr}} \frac{d}{dr} (\log\sqrt{-g_{00}}). \tag{11-10}$$

If now we replace $g_{rr}(r)$ by its value $\left[1 - \frac{2m(r)}{r}\right]^{-1}$, this differential equation for $\sqrt{-g_{00}}$ becomes

$$\frac{d}{dr} \log\sqrt{-g_{00}} = \frac{m(r) + 4\pi\kappa r^3 p_r(r)}{r(r - 2m(r))}. \tag{11-11}$$

Finally, consider the case of a perfect fluid, i.e., $p_r = p_\theta = p_\varphi = p$. From Equation 11-1,

$$-\frac{dp}{dr} = (\rho + p) \frac{d}{dr} \log\sqrt{-g_{00}}. \tag{11-12}$$

We may combine this with Equation 11-11 to give the Oppenheimer–Volkoff equation:

$$\frac{dp}{dr} = -\frac{(\rho + p)[m(r) + 4\pi\kappa r^3 p]}{r(r - 2m)}. \tag{11-13}$$

These are the principal equations of static, spherically symmetric fluid spheres. They are crucial for discussing relativistic stars. If we let U represent the classical Newtonian potential for a fluid ball, then Equation 11-11 replaces the classical law

$$\frac{dU}{dr} = \text{gravitational acceleration} = -\frac{m(r)}{r^2}, \tag{11-14}$$

corresponding to the fact that the shell of matter exterior to a point with radial coordinate r exerts no force on the material inside. Likewise Equations 11-12 and 11-13 replace the classical Newtonian law of hydrostatic equilibrium

$$\frac{dp}{dr} = \rho \frac{dU}{dr} = -\rho \frac{m(r)}{r^2}. \tag{11-15}$$

Note, however, that Equations 11-11, 11-12, and 11-13 alter these equations in several respects. Fluid pressure ordinarily prevents a body from collapsing due to gravitational force, and from Equation 11-14 this gravitational force is independent of the pressure. In Equation 11-11, however, we see that pressure also contributes to the gravitational force and thus further enhances gravitational collapse, which further increases the pressure, etc. (For these and other aspects of the relativistic equations of stellar structure, see Misner, Thorne, and Wheeler, 1973, Chapter 23.)

12

Cosmology

The Einstein Static Universe

To consider the universe on a cosmic scale, Einstein made a first approximation by smoothing out all local irregularities (planets, stars) into one cosmic fluid. One can imagine a fluid in which each "molecule" is an entire galaxy or group of galaxies. The simplest case would assume zero pressure, i.e., an incoherent dust; galaxies would attract each other but no pressure would result from collisions. From Equation 7-26, the world lines of the galaxies would be geodesics in space-time.

Can one always find, locally, a hypersurface V^3 orthogonal to the world lines? The answer, given by the Frobenius theorem (see Chapter 7) is that this is possible if and only if the vorticity vanishes:

$$d\nu\big|_{u^\perp} = -2\omega_{\alpha<\beta}\, dx^\alpha \wedge dx^\beta = 0.$$

This corresponds to fluid motion without spatial rotation.

Let us assume that one can find a single hypersurface V_0^3 that is orthogonal to the fluid world lines passing through it (in particular the vorticity vanishes at each point of V_0^3). Then by Gauss's lemma (an easy consequence of the first variation of energy formula, Eqn. 7-29), if we flow along each (geodesic) world line through V_0^3 for a proper time t, the resulting hypersurface V_t^3 is again orthogonal to the flow. By introducing local coordinates x^1, x^2, x^3 on V_0^3 and t along the world lines through V_0^3, we have constructed locally a "Gaussian" coordinate system (t, x^1, x^2, x^3). This coordinate system is *co-moving* in the sense that x^1, x^2, x^3 are constant along each world line. The metric in this local system is of the form

$$\begin{cases} ds^2 = -dt^2 + dl_t^2 \\ dl_t^2 = g_{\alpha\beta}(t,x)\,dx^\alpha dx^\beta \quad \text{the spatial metric on } V_t^3 \end{cases} \tag{12-1}$$

since the world lines are parameterized by proper time t and the spatial sections V_t^3 are always orthogonal to these world lines.

The same conclusions can be drawn if we allow a pressure that is *spatially constant,* that is,

$$dp\,\big|_{u^\perp} = 0$$

since again from Equation 7-26, the world lines are geodesics.

Einstein, in 1917, first attempted to construct a cosmological model with metric of the form of Equation 12-1 that was *static,* i.e., $g_{\alpha\beta}$ independent of t. This was a very reasonable assumption at the time since there was no evidence for the recession of the galaxies known today (Hubble's announcement was in 1929). A contradiction, however, arises immediately, for from the generalized Poisson equation (Eqn. 3-9) for a static universe

$$\nabla_V^2 \sqrt{-g_{00}} = 4\pi\kappa(\rho_0 + 3p)\sqrt{-g_{00}},$$

we have, since $g_{00} \equiv -1$ and $\rho_0 \geq 0$, $p \geq 0$, necessarily an empty universe!

Einstein made several attempts to remedy this situation. His most provocative (1917) involved changing the field equations by adding a term involving a new constant, the *cosmological constant* Λ:

$$R_{ij} - \frac{1}{2} g_{ij} R + \Lambda g_{ij} = 8\pi\kappa T_{ij}. \tag{12-2}$$

(Note that this is still consistent with the Cartan-Weyl-Lovelock result reported on p. 33.) This has the effect of replacing the stress-energy momentum tensor T_{ij} by $T_{ij} - (8\pi\kappa)^{-1}\Lambda g_{ij}$. This changes the Poisson equation to

$$\nabla_V^2 \sqrt{-g_{00}} = [4\pi\kappa(\rho_0 + 3p) - \Lambda]\sqrt{-g_{00}}$$

which in our case, $g_{00} = -1, p = 0$, yields

$$\Lambda = 4\pi\kappa\rho_0.$$

Einstein felt that it was natural to take ρ_0 = constant as a first approximation since the galaxies seemed to be uniformly distributed in space.

Note that the basic equation of static space, $R_{V^3} = 16\pi\kappa\rho_0$ (Eqn. 4-16), must be replaced, when using $\Lambda \neq 0$, by

$$R_{V^3} = 24\pi\kappa\rho_0.$$

Einstein took as his model, then, a static universe whose spatial sections V^3 were 3-spheres of constant curvature. If each V^3 has "radius" R, then $R_{V^3} = 6/R^2$, i.e., *the spatial universe has radius*

$$R = \frac{1}{\sqrt{4\pi\kappa\rho_0}} = \Lambda^{-1/2}.$$

Einstein felt that the introduction of $\Lambda \neq 0$ was a serious complication of his theory. Note that $\Lambda \neq 0$ forces even empty space $T_{ij} = 0$ to have curvature, though indeed Λ may be exceedingly small.

A striking advance was made by A. Friedmann in two papers (1922, 1924), in which he showed that one can put $\Lambda = 0$ *provided one allows the metric to vary in time*. We shall consider his model in some detail since it remains to this day the "standard" cosmological model.

The Friedmann Cosmology: Assumptions

Friedmann makes the following mathematical assumptions for his model.

(1) As Einstein before him, he assumes a model of the form $M^4 = R \times V^3$ but it is *not* assumed that M^4 carries the product metric.

(2) It is assumed that the spatial model space V^3 carries again a *metric of constant curvature;* by normalization we may assume this curvature K_V is 1, 0, or -1.

(3) It is assumed that the metric on M^4 is of the form

$$ds^2 = -dt + G^2(t)dl^2,$$

where dl^2 is the metric on V^3 and $G^2(t)$ is a function to be determined. Thus the metric on the spatial section V_t^3, $dl_t^2 = G^2(t)dl^2$, is conformally equivalent to that on V^3. In particular, *V_t^3 is a space of constant curvature* $K_t = G^{-2}(t)K_V$ (think, for example, of a sphere of radius $G(t)$ in Euclidean space).

Some justification for these assumptions is in order. If we again assume that the cosmic fluid is without vorticity (no rotation) and has spatially constant pressure p, then the world lines of the fluid are again geodesics. Local (Gaussian) coordinates can be introduced such that the metric assumes the local form

$$ds^2 = -dt^2 + g_{\alpha\beta}(t, x)\, dx^\alpha\, dx^\beta. \tag{12-3}$$

Condition (1) assumes that globally M^4 decomposes into these time lines and global orthogonal spatial sections. Note that in a metric of the form of Equation 12-3 the time lines are, conversely, automatically geodesics. In fact, such lines are clearly locally the "longest" curves joining a nearby pair of points on a time line.

As for condition (2), it has long been noted that the galaxies seem to be *isotropically* distributed with respect to the earth, i.e., their distribution is (approximately) uniform in all directions. Also, it is unreasonable to assume that the earth occupies a special position in the universe. We therefore assume that the spatial universe V_t^3 is always *locally isotropic*. By this we mean that each $P \in V_t^3$ has a neighborhood in V_t^3 that is invariant (leaving P fixed) under a local three-parameter group of isometries (isomorphic to the orthogonal group $O(3)$). In particular, the Riemann sectional curvature K_t (a function $\mathsf{K}_t(\Pi_p)$ of tangent 2-planes at P) is in fact independent of the plane Π_p at P since $O(3)$ is transitive on 2-planes, $\mathsf{K}_t(\Pi_p) = K_t(P)$. It is reasonable to expect that $K_t(P)$ is also independent of the point P; it is interesting that this already follows from local isotropy via the classical

Schur's theorem If V^n, $n \geq 3$, is Riemannian and locally isotropic, $\mathsf{K}(\Pi_p) = K(P)$, then K is actually constant (we then say that V^n has constant (sectional) curvature).

PROOF If V^n is locally isotropic and if ξ is any unit vector at P, then $\mathrm{Ric}(\xi, \xi) = (n - 1)K(P)$, since $\mathrm{Ric}(\xi, \xi)$ is a sum of curvatures for $(n - 1)$ 2-planes at P (see Eqn. 4-5). Hence the Ricci tensor is proportional to the metric tensor $R_{\alpha\beta} = (n - 1)K(P)g_{\alpha\beta}$. The scalar curvature is then $R = n \cdot (n - 1)K(P)$, and, consequently, in local coordinates $\partial R/\partial x^\alpha = n(n - 1)\partial K/\partial x^\alpha$. On the other hand, from Equation 7-22, $\partial R/\partial x^\alpha = 2R^\beta{}_{\alpha;\beta} = 2(n - 1)(K\delta^\beta{}_\alpha)_{;\beta} = 2(n - 1) \cdot \partial K/\partial x^\alpha$. Thus if $n \neq 2$, $\partial K/\partial x^\alpha = 0$, as desired.

It is also a mathematical fact that a space of constant curvature is, conversely, locally isotropic. It need not be isotropic in the large, by which we mean that the local isometries need not be extendable to the entire manifold. For example, consider the flat ($K = 0$) two-dimensional torus T^2, considered as R^2 with identifications $(x, y) \sim (x + n, y + m)$, n, m, integers (Figure 12-1). The rotation (say, through 45°) about P is an isometry of a small disc centered at P, but this isometry cannot be extended to a disc so large as to include the equivalent point $P' \sim P$. If it did, the isometry would have to leave P fixed (as it does in the small disc) and yet send $P \sim P'$ into $Q \not\sim P'$. Directions through P are locally, but not globally, equivalent. This is also evident from the fact that the closed geodesic $\overline{PP'}$ has length 1 while the closed geodesic $\overline{PP''}$ has length $\sqrt{2}$.

Conditions (2) and (3) offer a very simple scheme for insuring the local isotropy of the spatial sections V_t^3. *They allow the metric in the spatial sections to change in time but always isotropically.*

The condition of local isotropy (constant curvature) is the most severe restriction of a differential geometric nature that can be imposed on a manifold V^n. By a constant change of distance scale the curvature may be assumed to be $+1$, 0, or -1. Models for these three cases are the unit n-sphere $S^n \subset R^{n+1}$ (with $K = 1$), R^n itself ($K = 0$) and the "hyperbolic

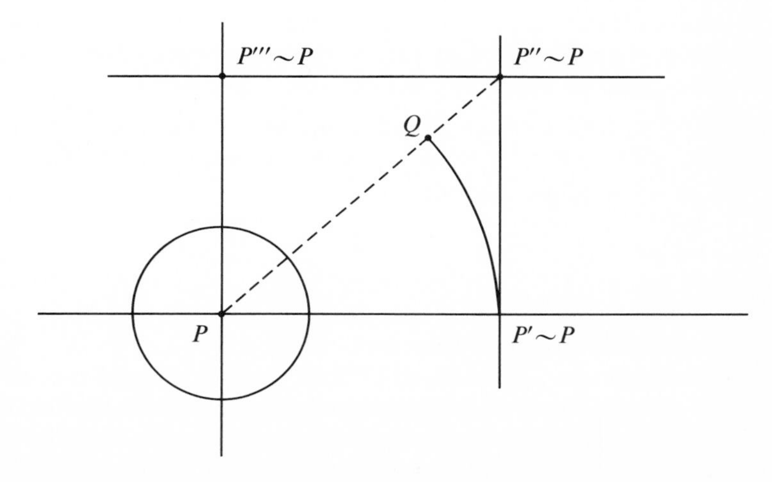

FIGURE 12-1

n-space" $H^n(K = -1)$. These models are actually globally isotropic. Any complete V^n of constant curvature $+1$, 0, or -1 has the corresponding model space as universal covering manifold. Those V^n with $K = 1$ are compact, those with zero or negative curvature may be compact or not compact (for example, the n-torus T^n is compact). A *compact* Riemannian V^n of negative curvature cannot be globally isotropic; in fact, Bochner's theorem states that a compact V^n with negative definite Ricci tensor admits no globally defined one-parameter group of isometries (and therefore no Killing vector fields). (For proof see p. 159ff.)

Finally, we summarize the *physical* assumptions of the Friedmann cosmological model. The cosmic fluid is assumed a perfect fluid with spatially constant density and pressure, $\rho = \rho(t), p = p(t)$, whose world lines are the geodesic time lines.

The Friedmann Cosmology: The Solution

We are now to determine $G(t)$ so that the resulting space-time M^4 satisfies the original Einstein equations. Note that the contradiction that arose in Einstein's static model does not apply here since the generalized Poisson equation (Eqn. 3-9) does not hold in a nonstatic universe.

A space of constant sectional curvature has a metric with a well-known local canonical form, but we shall have no use for this expression. Instead we shall rely on simple consequences of local isotropy.

Consider Einstein's equation (Eqn. 4-12):

$$8\pi\kappa\rho(t) = 8\pi\kappa \mathrm{T}\left(\frac{\partial}{\partial t}, \frac{\partial}{\partial t}\right) = \frac{1}{2} R_{V_t^3} + \kappa_1\kappa_2 + \kappa_1\kappa_3 + \kappa_2\kappa_3. \tag{12-4}$$

First, evaluate $R_{V_t^3}$. As we have already remarked, V_t^3 has constant curvature $G^{-2}(t)K_V$ (where $K_V = 1, 0$, or -1 is the curvature of V^3). Since $R_{V_t^3}$ is the sum of six sectional curvatures,

$$\frac{1}{2} R_{V_t^3} = \frac{3K_V}{G^2(t)}. \tag{12-5}$$

We next need the principal curvatures of $V_t^3 \subset M^4$. Let X_1, X_2, X_3 be orthonormal vectors at a point P_V of V^3. Lift them to each of the spatial sections V_t^3 (keeping their spatial components unchanged) and call the resulting fields X again (Figure 12-2). They are still orthogonal in the dl_t^2 metric but they are not unit vectors. Define the unit vectors $e_\alpha = G^{-1}(t)X_\alpha$, $\alpha = 1, 2, 3$. Of course, they are orthogonal to the unit vector

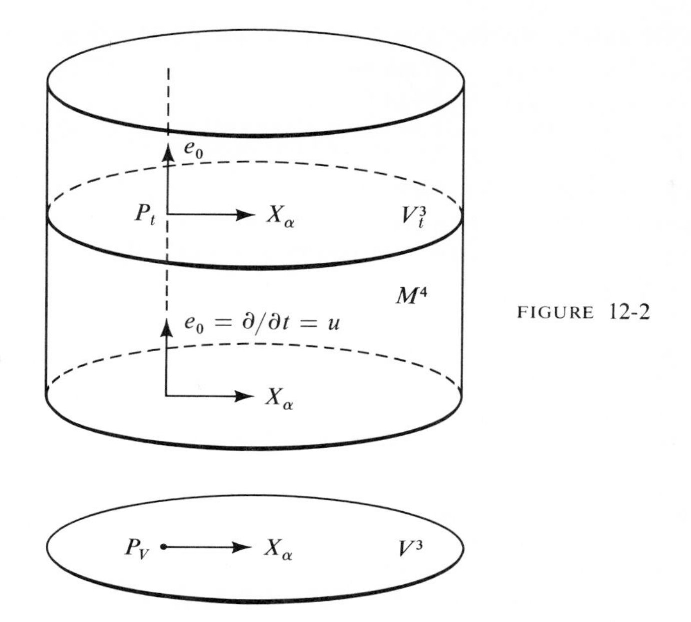

FIGURE 12-2

$e_0 = \dfrac{\partial}{\partial t} = u$ (the unit 4-velocity of the cosmic fluid). Since the X_α are, by construction, invariant under time translation, $[e_0, X] = \nabla_{e_0} X_\alpha - \nabla_{X_\alpha} e_0 = 0$. For the second fundamental form of $V_t^3 \subset M^4$ we have

$$\begin{aligned} \mathrm{b}(X_\alpha) &= -\nabla_{X_\alpha} e_0 = -\nabla_{e_0} X_\alpha = -\nabla_{e_0}(G \cdot e_\alpha) \\ &= -\frac{\partial G}{\partial t} e_\alpha - G \cdot \nabla_{e_0} e_\alpha. \end{aligned}$$

Now $\mathrm{b}(X_\alpha)$ is tangent to V_t^3, hence $G \nabla_{e_0} e_\alpha$ is also. But $\nabla_{e_0} e_\alpha$ is also orthogonal to e_α. We claim that $\nabla_{e_0} e_\alpha = 0$.

PROOF Let P_t be the event in M^4 over $P_V \in V^3$ with time coordinate t. By local isotropy there exists an isometry (rotation) of some neighborhood of P_V whose differential has the line through the given X_α as fixed set. This isometry lifts to an isometry φ of a neighborhood of the entire t-line through P_t. The differential φ_* along this t-line has fixed set spanned by e_α and e_0. Let $\mathscr{T}_h$ be parallel translation along the t-line from P_{t+h} to P_t. Then

$$\nabla_{e_0} e_\alpha = \lim_{h \to 0} \frac{\mathscr{T}_h e_\alpha(P_{t+h}) - e_\alpha(P_t)}{h}.$$

Since φ is an isometry, $\varphi_* \circ \mathcal{T}_h = \mathcal{T}_h \circ \varphi_*$, and consequently

$$\varphi_* \nabla_{e_0} e_\alpha = \nabla_{e_0} e_\alpha .$$

Thus $\nabla_{e_0} e_\alpha$ lies in the plane spanned by e_α and e_0 and yet is orthogonal to both, $\nabla_{e_0} e_\alpha = 0$. (The following briefer argument is compelling: since all directions in space are "equivalent," the only direction that $e_\alpha \to \nabla_{e_0} e_\alpha$ can single out is the line through e_α!)

Thus

$$\mathsf{b}(X_\alpha) = -\frac{\partial G}{\partial t} e_\alpha = -\frac{1}{G}\frac{\partial G}{\partial t} X_\alpha$$

shows that the principal curvatures of V_t^3 are

$$\kappa_1 = \kappa_2 = \kappa_3 = -\frac{\dot{G}}{G}, \tag{12-6}$$

where the dot refers to differentiation with respect to time. Einstein's equation (Eqn. 12-4) together with Equations 12-5 and 12-6 then give the Friedmann equation:

$$\boxed{\dot{G}^2(t) + K_V = \frac{8}{3}\pi\kappa\rho(t)G^2(t).} \tag{12-7}$$

How are $G(t)$ and $\rho(t)$ related? The mass 3-form on each V_t^3 is $\rho\omega_V^3$. From the equations of motion, $\mathrm{div}(\rho u) = -p \ \mathrm{div}\, u$ (Eqn. 7-25), where again $u = e_0$, and, in particular from Equation 7-28, where $\omega_\perp^3 = \omega_V^3$, we see that mass is conserved in the case of $p = 0$, i.e., for an incoherent dust. We add the assumption (Friedmann) that *the cosmic fluid is a dust*, $p \equiv 0$, used previously by Einstein. We conclude that mass is conserved, that is,

$$\rho(t)G^3(t) \text{ is constant,}$$

since volumes vary as the cube of $G(t)$. Thus if we define the constant m by

$$m \equiv \frac{8}{3}\pi\kappa\rho(t)G^3(t) = \text{constant},$$

the Friedmann equation becomes, in the case of dust,

$$\dot{G}^2(t) + K_V = \frac{m}{G(t)}. \tag{12-8}$$

We shall consider separately each of the three cases $K_V = 1, 0$, and -1.

CASE I. $K_V = 0$. *Space is flat*, $\dot{G}^2 = m/G$, and so $G(t) = [^3/_2 \sqrt{m}\, t + \text{constant}]^{2/3}$. This model is called the *Einstein–de Sitter Universe*.

CASE II. $K_V = +1$. *Space is positively curved.* If V^3 is complete then V^3 is covered by a 3-sphere. Thus each spatial section V_t^3 is, if complete, compact, i.e., each V_t^3 is a "closed" spatial universe. $\dot{G}^2 = m/G - 1 = (m - G)/G$, and so the differential equation for G is

$$\frac{\sqrt{G}}{\sqrt{m - G}}\, dG = dt.$$

Make the substitution. $G = m \sin^2 \eta$ to get the parametric solution

$$\begin{cases} G(\eta) = \dfrac{m}{2}(1 - \cos 2\eta) \\ \\ t(\eta) = \dfrac{m}{2}(2\eta - \sin 2\eta) + \text{constant}, \end{cases}$$

the parametric equations of a *cycloid*. There is also apparently a singular solution, $G(t) \equiv m = \text{constant}$, but use of even a portion of this is ruled out by the same argument as used in the Einstein static universe (Poisson's equation, Eqn. 3-9).

CASE III. $K_V = -1$. *Space is negatively curved*, $\dot{G}^2 = m/G + 1 \geq 1$ with solution

$$\begin{cases} G(\eta) = \dfrac{m}{2}(\cosh(2\eta) - 1) \\ \\ t(\eta) = \dfrac{m}{2}(\sinh(2\eta) - 2\eta) + \text{constant}. \end{cases}$$

In all these cases $G(t_0) = 0$ signifies a singularity at time t_0; if $\mathcal{U}$ is a compact set on V^3 then $\text{vol}(\mathcal{U}_{t_0}) = 0$. Space explosively expands starting at this $t = t_0$ (the "big bang"). Plot these three cases, adjusting t so that $t_0 = 0$ (Figure 12-3).

In summary, all models start from a singularity. The flat and negatively curved cases expand forever, while the positively curved space reaches a maximum expansion $G = m$, and then contracts back into the singular

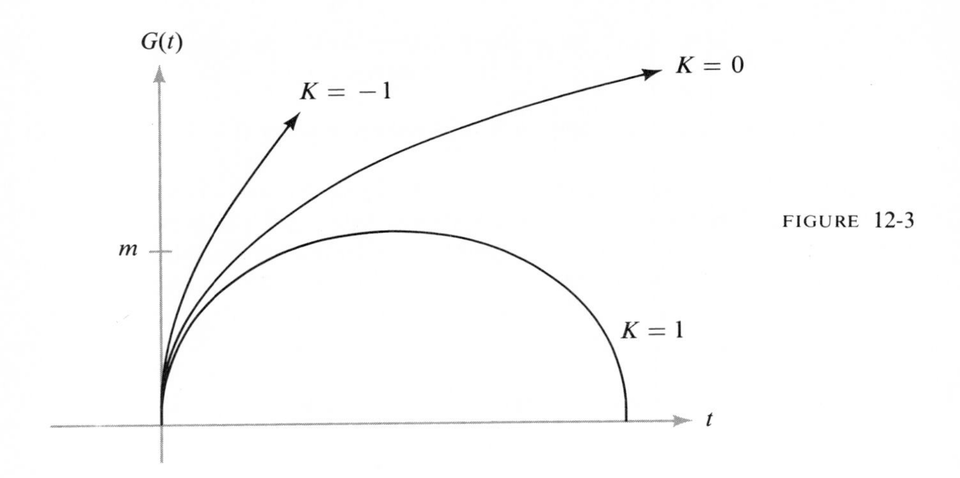

FIGURE 12-3

state. It is remarkable that Friedmann's dynamic universe models were introduced about seven years before astronomers considered an expanding universe! In 1929 Hubble showed that, on the average, galaxies are receding from our own galaxy at a rate that is proportional to their distance from us. If we imagine galaxies distributed uniformly on a 3-sphere of radius $G(t)$ with our galaxy at the north pole ($\alpha = 0$), then a galaxy with colatitude α will recede from us at speed $\dot{G}(t)\alpha$, if we neglect the local changes $\dot{\alpha}$. Thus the *Hubble parameter,*

$$h(t) = \frac{\dot{G}(t)}{G(t)} \tag{12-9}$$

representing the logarithmic rate of change of distance from us, will be approximately independent of the galaxy, though it might depend on time. This is what was observed by Hubble and convinced astronomers of the "expanding universe." We will have a few words to say about the singularity later on in this chapter.

Returning to the Friedmann models, we get from Equations 12-9 and 12-7 that $h^2 + K_V/G^2 = \frac{8}{3}\pi\kappa\rho$, i.e.,

$$\begin{cases} K_V = \tfrac{1}{3} \cdot G^2(t)\Delta(t) \\ \Delta = 8\pi\kappa\rho - 3h^2 \end{cases} \tag{12-10}$$

and so

$$\begin{cases} \Delta > 0 \rightarrow (K_V = 1) \rightarrow \text{compactness, and expansion, then contraction} \\ \Delta \leq 0 \rightarrow (K_V = 0 \text{ or } -1) \rightarrow \text{expansion forever.} \end{cases} \tag{12-11}$$

Thus a knowledge of the mean density of matter ρ in the universe and a knowledge of Hubble's h (via the observed recession of the galaxies) determines the fate of the universe in this Friedmann model. As of today, observations do not seem to yield a sign for the quantity Δ. (How much matter between galaxies is invisible? How does one measure distances to determine h?)

Two remarks are in order. First, we have derived the Friedmann models by merely solving the single Einstein equation (Eqn. 12-4), that is, the equation

$$8\pi\kappa T_{00} = R_{00} + \frac{1}{2}R.$$

It is not difficult to see, however, using local isotropy, that the remaining Einstein equations are also satisfied.

Second, the Friedmann condition of constant curvature (local isotropy) clearly does not hold exactly in our universe. We must study how relaxing this most stringent condition will effect the Friedmann conclusions. Will the sign of the curvature still dictate expansion and contraction? Will the density and recession rate still influence compactness? Will there always be singularities? We turn now to these questions.

The (Landau-)Raychaudhuri Equation

Consider a time-like unit vector field u in a pseudo-Riemannian M^4. We shall compute the space-time divergence of the acceleration vector $\nabla_u u$:

$$\begin{aligned} \operatorname{div}(\nabla_u u) &= (u^i{}_{;k}u^k)_{;i} = u^i{}_{;ki}u^k + u^i{}_{;k}u^k{}_{;i} \\ &= u^i{}_{;ki}u^k + u_{i;k}u^{k;i}. \end{aligned}$$

From the Ricci identities

$$u^i{}_{;ki} - u^i{}_{;ik} = u^j R^i{}_{jik} = u^j R_{jk},$$

and so

$$\mathrm{div}(\nabla_u u) = u^i{}_{;ik}u^k + u^j R_{jk} u^k + u_{i;k}u^{k;i},$$

that is,

$$\mathrm{div}(\nabla_u u) = u(\mathrm{div}\, u) + \mathrm{Ric}(u, u) + u_{i;k}u^{k;i}.$$

Now decompose $u_{i;k}$ into symmetric and antisymmetric parts

$$\begin{cases} u_{i;k} = \tilde{\theta}_{ik} + \tilde{\omega}_{ik} \\ \tilde{\theta}_{ik} = \dfrac{1}{2}(u_{i;k} + u_{k;i}) \\ \tilde{\omega}_{ik} = \dfrac{1}{2}(u_{i;k} - u_{k;i}). \end{cases}$$

Then

$$\mathrm{div}(\nabla_u u) = u(\mathrm{div}\, u) + \mathrm{Ric}(u, u) + \tilde{\theta}_{ik}\tilde{\theta}^{ik} - \tilde{\omega}_{ik}\tilde{\omega}^{ik}. \qquad (12\text{-}12)$$

Given an event $p \in M^4$, let e_1, e_2, e_3 be locally defined vector fields near p satisfying $[u, e_\alpha] = 0$, that is, e_α are invariant under the flow generated by u, and $u = e_0, e_1, e_2, e_3$ are orthonormal at p. We may even assume that u, e_1, e_2, e_3 are coordinate vectors, $e_i = \partial/\partial x^i$, for some coordinate system near p. Then, at p, we have

$$\nabla_{e_i} u = u^j{}_{;i} e_j \text{ and } \langle \nabla_{e_i} u, e_k \rangle = u_{k;i},$$

that is,

$$\begin{cases} \tilde{\theta}_{ij} = \dfrac{1}{2}(\langle \nabla_{e_j} u, e_i \rangle + \langle \nabla_{e_i} u, e_j \rangle) \\ \tilde{\omega}_{ij} = \dfrac{1}{2}(\langle \nabla_{e_j} u, e_i \rangle - \langle \nabla_{e_i} u, e_j \rangle) \end{cases}$$

which is merely the four-dimensional version of Equation 7-7. We relate these forms to those of Equation 7-7. Note

$$\tilde{\theta}_{ij}\tilde{\theta}^{ij} = \tilde{\theta}_{00}\tilde{\theta}^{00} + 2\tilde{\theta}_{0\beta}\tilde{\theta}^{0\beta} + \tilde{\theta}_{\alpha\beta}\tilde{\theta}^{\alpha\beta}$$

and

$$\tilde{\theta}_{00} = \frac{1}{2}(\langle \nabla_u u, u \rangle + \langle u, \nabla_u u \rangle) = 0$$

$$\tilde{\theta}_{0\beta} = \frac{1}{2}(\langle \nabla_{e_\beta} u, u \rangle + \langle \nabla_u u, e_\beta \rangle)$$

$$= \frac{1}{2}\langle \nabla_u u, e_\beta \rangle = \frac{1}{2} k_\beta,$$

where $\nabla_u u = \sum_{\beta=1}^{3} k_\beta e_\beta$ is the curvature vector. A simple calculation with our orthonormal frame gives $\tilde{\theta}^{0\beta} = -\tilde{\theta}_{0\beta}$ and $\tilde{\theta}^{\alpha\beta} = \tilde{\theta}_{\alpha\beta}$, and so

$$\tilde{\theta}_{ij}\tilde{\theta}^{ij} = -\frac{1}{2}\sum k_\beta^2 + \tilde{\theta}_{\alpha\beta}\tilde{\theta}^{\alpha\beta}. \tag{12-13}$$

Likewise,

$$\tilde{\omega}_{ij}\tilde{\omega}^{ij} = -\frac{1}{2}\sum k_\beta^2 + \tilde{\omega}_{\alpha\beta}\tilde{\omega}^{\alpha\beta}. \tag{12-14}$$

If we let $\Theta = (\theta_{\alpha\beta})$ and $\Omega = (\omega_{\alpha\beta})$ be, as in Chapter 7, the restrictions of $\tilde{\Theta}$ and $\tilde{\Omega}$ to $u^\perp$, that is, $\theta_{\alpha\beta} = \tilde{\theta}_{\alpha\beta}$, $\omega_{\alpha\beta} = \tilde{\omega}_{\alpha\beta}$, we then get, from Equations 12-12, 12-13, and 12-14 and 7-16 the so-called *(Landau)-Raychaudhuri equation*

$$\frac{d\theta}{d\tau} = -\mathrm{Ric}(u, u) - \theta_{\alpha\beta}\theta^{\alpha\beta} + \omega_{\alpha\beta}\omega^{\alpha\beta} + \mathrm{div}(\nabla_u u), \tag{12-15}$$

since $u(\theta) = d\theta/d\tau$, where τ is proper time along the flow lines of u. The geometric significance of the forms $(\theta_{\alpha\beta})$ and $(\omega_{\alpha\beta})$ has already been discussed on p. 76.

In Equation 12-15 it is usual to further decompose $\theta_{\alpha\beta}$ by means of the *rate of shear tensor* (Eqn. 7-18), at the given event p,

$$\sigma_{\alpha\beta} = \theta_{\alpha\beta} - \frac{1}{3}\theta\delta_{\alpha\beta}.$$

Then

$$\theta_{\alpha\beta}\theta^{\alpha\beta} = \sigma_{\alpha\beta}\sigma^{\alpha\beta} + \frac{1}{3}\theta^2$$

and the Raychaudhuri equation becomes

$$\frac{d\theta}{d\tau} = -\mathsf{Ric}(u, u) - \sigma_{\alpha\beta}\sigma^{\alpha\beta} - \frac{1}{3}\theta^2 + \omega_{\alpha\beta}\omega^{\alpha\beta} + \mathsf{div}(\nabla_u u). \quad (12\text{-}16)$$

The Geometry of a Vorticity-Free Flow

Consider a flow in M^4 that is vorticity-free, i.e., $\omega_{\alpha\beta}\omega^{\alpha\beta} = 0$. By the Frobenius theorem (p. 77ff) we know that locally there are hypersurfaces that are everywhere orthogonal to the flow lines, i.e., locally M^4 is of the form $R \times V^3$ with metric

$$\begin{cases} ds^2 = g_{00}(t, x)\, dt^2 + dl_t^2 \\ dl_t^2 = g_{\alpha\beta}(t, x)\, dx^\alpha\, dx^\beta, \text{ the metric for } V_t^3. \end{cases} \quad (12\text{-}17)$$

The velocity vector $u = (-g_{00})^{-1/2}\partial/\partial t$ is the unit normal field to the spatial sections V_t^3, where t = constant. Recall that the second fundamental form b for the hypersurface V_t^3 is defined by $\mathsf{b}(X) = -\nabla_X u$, and consequently the symmetric bilinear form B associated with b satisfies

$$\mathsf{B}(X, Y) = \langle \mathsf{b}(X), Y\rangle = -\langle \nabla_X u, Y\rangle.$$

Thus, from Equation 7-6

$$b_{\alpha\beta} = \mathsf{B}\left(\frac{\partial}{\partial x^\alpha}, \frac{\partial}{\partial x^\beta}\right) = -\theta_{\alpha\beta}, \quad (12\text{-}18)$$

that is, the rate of strain tensor $(\theta_{\alpha\beta})$ is the negative of the second fundamental form $(b_{\alpha\beta})$! We also have

$$\theta = \mathsf{div}\, u = \mathsf{tr}(\Theta) = -\mathsf{tr}(\mathsf{b}) = -H, \quad (12\text{-}19)$$

the expansion is the negative of the mean curvature of V_t^3.

A Generalized Poisson Equation for Vorticity-Free Flows

We now return to an aspect of general relativity that formed the cornerstone of our entire development of this subject, namely the generalized Poisson equation. We shall need a generalization of Levi-Civita's equation (Eqn. 3-3) to nonstatic universes. We shall assume a local metric description as in Equation 12-17, i.e., we assume the vorticity vanishes, and our goal is to express the spatial Laplacian of $\sqrt{-g_{00}}$ in terms of the geometry of V_t^3 and the stress-energy-momentum tensor of M^4. Introduce the following notation ($\varphi = \sqrt{-g_{00}}$ is to remind us of a scalar potential on V_t^3):

$$\begin{cases} \varphi = \sqrt{-g_{00}} \\ \mathsf{div}_V = \text{the spatial divergence operator on } V_t^3 \\ \mathsf{grad}_V = \text{the spatial gradient operator on } V_t^3 \\ \nabla_V^2 = \text{the spatial Laplace operator.} \end{cases}$$

Let e_1, e_2, e_3 be orthonormal tangent vectors to V_t^3 at p; then $u = e_0, e_1, e_2, e_3$ are orthonormal at p and from Equation 7-15

$$\mathsf{div}(\nabla_u u) = -\langle \nabla_u(\nabla_u u), u \rangle + \sum_{\alpha=1}^{3} \langle \nabla_{e_\alpha}(\nabla_u u), e_\alpha \rangle.$$

Since $\nabla_u u$ is tangent to V_t^3, we may write this as

$$\mathsf{div}(\nabla_u u) = -\langle \nabla_u(\nabla_u u), u \rangle + \mathsf{div}_V(\nabla_u u). \tag{12-20}$$

Now

$$\langle \nabla_u(\nabla_u u), u \rangle = u\langle \nabla_u u, u \rangle - \langle \nabla_u u, \nabla_u u \rangle = -\|\nabla_u u\|^2, \tag{12-21}$$

and from the lemma on page 129,

$$\begin{aligned} \mathsf{div}_V(\nabla_u u) &= \mathsf{div}_V(\varphi^{-1}\mathsf{grad}_V\varphi) \\ &= -\varphi^{-2}\|\mathsf{grad}_V\varphi\|^2 + \varphi^{-1}\nabla_V^2\varphi \\ &= -\|\nabla_u u\|^2 + \varphi^{-1}\nabla_V^2\varphi, \end{aligned}$$

and so from Equations 12-20 and 12-21

$$\mathsf{div}(\nabla_u u) = \varphi^{-1}\nabla_V^2\varphi. \tag{12-22}$$

Substitute this into the Raychaudhuri equation (Eqn. 12-15) using $\theta_{\alpha\beta}\theta^{\alpha\beta} = b_{\alpha\beta}b^{\alpha\beta} = \mathrm{tr}(\mathsf{b}^2)$ and $\theta = -H$,

$$\varphi^{-1}\nabla_V^2\varphi = \mathsf{Ric}(u, u) + \mathrm{tr}(\mathsf{b}^2) - \frac{dH}{d\tau}.$$

Along the world line, we have $dH/d\tau = \varphi^{-1}\partial H/\partial t$, and we have an equation first derived (in the Riemannian case) by Duschek,

$$\nabla_V^2 \varphi = \mathrm{Ric}(u, u)\varphi + \mathrm{tr}(\mathsf{b}^2)\varphi - \frac{\partial H}{\partial t}. \tag{12-23}$$

Now $\mathrm{Ric}(u, u) = \mathrm{Ric}(\varphi^{-1}\partial/\partial t, \varphi^{-1}\partial/\partial t) = (-g_{00})^{-1}R_{00} = -g^{00}R_{00} = -R_0^0$. Using Einstein's equation (Eqn. 3-11), Equation 12-23 yields the *generalized Poisson equation*

$$\nabla_V^2 \sqrt{-g_{00}} = 4\pi\kappa[T_1^1 + T_2^2 + T_3^3 - T_0^0]\sqrt{-g_{00}} + \mathrm{tr}(\mathsf{b}^2)\sqrt{-g_{00}} - \frac{\partial H}{\partial t}. \tag{12-24}$$

In the static case V_t^3 is totally geodesic, $H = 0 = \mathrm{tr}(\mathsf{b}^2)$, and Equations 12-23 and 12-24 reduce to the Levi-Civita equations of Chapter 3.

It is interesting to note the similarity with electromagnetic theory. In electrostatics, in Euclidean space, the electrostatic potential φ satisfies Poisson's equation $\nabla^2\varphi = 4\pi\sigma$, while in electrodynamics we have (see p. 123ff)

$$\nabla^2\varphi = 4\pi\sigma + \frac{\partial}{\partial t}\,\mathrm{div}\,\mathbf{A}. \tag{12-25}$$

Thus, comparing with Equation 12-24 we see that the gravitational terms involving the second fundamental form b of V_t^3 play a role suggestive of that of the vector potential $\mathbf{A}$ in electrodynamics.

Return now to the Friedmann cosmological model. Note that from Equation 12-6 we have

$$H = \mathrm{tr}(\mathsf{b}) = \kappa_1 + \kappa_2 + \kappa_3 = -\frac{3\dot{G}}{G} = -3h, \tag{12-26}$$

where h is again the Hubble parameter. Also, considering the case where p might not vanish, $\mathrm{Ric}(u, u) = 4\pi\kappa(\rho + 3p)$. Since $g_{00} = -1$, the Poisson equation (Eqn. 12-24) becomes

$$0 = 4\pi\kappa(\rho + 3p) + 3\frac{\dot{G}^2}{G^2} + 3\frac{d}{dt}\left(\frac{\dot{G}}{G}\right),$$

and we get the second Friedmann equation

$$\boxed{\ddot{G} = -\frac{4}{3}\pi\kappa(\rho + 3p)G} \tag{12-27}$$

Corollary $\ddot{G} < 0$ for all Friedmann models with $\rho + 3p > 0$. Thus the expansion is always *decelerating* in these models, (see Figure 12-3). This again rules out the singular solution G = constant of Case II, page 147.

General Vorticity-Free Cosmologies: The Influence of Curvature on Expansion

Consider again the metric of Equation 12-17 associated with a vorticity-free model with unit velocity vector u:

$$\begin{cases} ds^2 = g_{00}(t, x)\, dt^2 + dl_t^2 \\ dl_t^2 = g_{\alpha\beta}(t, x)\, dx^\alpha dx^\beta \quad \text{the metric on } V_t^3. \end{cases}$$

Let

$$\omega_V^3 = i_u\omega^4 = \sqrt{g_V}\, dx^1 \wedge dx^2 \wedge dx^3$$

be the volume form for V_t^3, $g_V = \det(g_{\alpha\beta})$. If $\mathcal{U}_t$ is a compact subset of V_t^3, its volume is

$$\text{vol}(t) = \text{vol}(\mathcal{U}_t) = \int_{\mathcal{U}_t} \omega_V^3.$$

Let, as usual

$$\begin{cases} \varphi = \sqrt{-g_{00}} \\ u = \varphi^{-1}\dfrac{\partial}{\partial t}. \end{cases}$$

Then, for volume variation under time translation, as in Equation 7-14,

$$\begin{aligned} \text{vol}'(t) &= \int_{\mathcal{U}_t} \mathcal{L}_{\frac{\partial}{\partial t}}\, \omega_V^3 = \int_{\mathcal{U}_t} i_{\frac{\partial}{\partial t}}\, d\omega_V^3 = \int_{\mathcal{U}_t} i_{\frac{\partial}{\partial t}}\, d i_u \omega^4 \\ &= \int_{\mathcal{U}_t} i_{\frac{\partial}{\partial t}}\, (\text{div}\, u)\omega^4 = \int_{\mathcal{U}_t} \varphi\, \text{div}\, u\, \omega_V^3, \end{aligned}$$

which, from Equation 12-19, yields the classic formula

$$\mathsf{vol}'(t) = -\int_{\mathcal{U}_t} \varphi \cdot H\omega_V^3, \tag{12-28}$$

where H is the mean curvature of the spatial section V_t^3. Thus the volume of the region $\mathcal{U}_t$ will be increasing if $H < 0$ there. But from Einstein's equations, where (κ_α) are principal curvatures of V_t^3,

$$\begin{aligned} 16\pi\kappa\rho &= 16\pi\kappa\mathsf{T}(u, u) = R_{V_t^3} + 2\,\mathrm{tr}(\mathsf{b}\wedge\mathsf{b}) \\ &= R_{V_t^3} + 2(\kappa_1\kappa_2 + \kappa_1\kappa_3 + \kappa_2\kappa_3) \\ &= R_{V_t^3} + (\kappa_1 + \kappa_2 + \kappa_3)^2 - (\kappa_1^2 + \kappa_2^2 + \kappa_3^2), \end{aligned}$$

that is,

$$16\pi\kappa\rho = R_{V_t^3} + H^2 - \mathrm{tr}(\mathsf{b}^2). \tag{12-29}$$

We then have, as a consequence of Equation 12-28 and Einstein's equations:

> **Corollary** (MacCallum, 1971). If $R_{V_t^3} \leq 0$ in $\mathcal{U}_t$ for all $t \in [0, b]$, and if $\mathcal{U}_t$ is expanding in volume at $t = 0$, then H cannot change sign and consequently $\mathsf{vol}(\mathcal{U}_t)$ is increasing for $t \in [0, b]$.

This is illustrated in the Friedmann (isotropic) model, Case III, page 147.

For more information about volume expansion we compute the second variation. From Equation 12-28

$$\mathsf{vol}''(t) = -\int_{\mathcal{U}_t} \frac{\partial\varphi}{\partial t} H\omega_V^3 - \int_{\mathcal{U}_t} \varphi\frac{\partial H}{\partial t}\omega_V^3 + \int_{\mathcal{U}_t} \varphi^2 H^2\omega_V^3,$$

and using Equation 12-23 we get Duschek's formula (1936)

$$\begin{aligned} \mathsf{vol}''(t) = &\int_{\mathcal{U}_t} \varphi\nabla_V^2\varphi\omega_V^3 - \int_{\mathcal{U}_t} H\frac{\partial\varphi}{\partial t}\omega_V^3 \\ &+ \int_{\mathcal{U}_t} \varphi^2[H^2 - \mathrm{tr}(\mathsf{b}^2) - \mathsf{Ric}(u, u)]\omega_V^3. \end{aligned} \tag{12-30}$$

Using again $\mathrm{tr}(\mathsf{b}^2) = H^2 - 2\mathrm{tr}(\mathsf{b}\wedge\mathsf{b})$ and the geometry of the Einstein tensor

$$\frac{1}{2} R_{V_t^3} + \mathrm{tr}(\mathsf{b} \wedge \mathsf{b}) = \mathsf{G}(u, u) = \mathsf{Ric}(u, u) + \frac{1}{2} R,$$

where as usual R is the scalar curvature of M^4, we get from Equation 12-30

$$\mathrm{vol}''(t) = \int_{\mathcal{U}_t} \varphi \nabla_V^2 \varphi \omega_V^3 - \int_{\mathcal{U}_t} H \frac{\partial \varphi}{\partial t} \omega_V^3 + \int_{\mathcal{U}_t} \varphi^2 [\mathsf{Ric}(u, u) + R - R_V] \omega_V^3. \tag{12-31}$$

Consider, for example, a vorticity-free incoherent dust (the flow lines are geodesics and there are hypersurfaces orthogonal to these flow lines). In terms of the Gaussian coordinate system with V_t^3 orthogonal to the flow (a so-called co-moving coordinate system) the metric is of the form in Equation 12-17 with $g_{00} = -1$, since proper time is employed along the world lines. We also have $\mathsf{Ric}(u, u) = 4\pi\kappa(\rho + 3p) = 4\pi\kappa\rho$ and $R = -8\pi\kappa T = 8\pi\kappa(\rho - 3p)$, i.e., $R = 8\pi\kappa\rho$. By conservation of mass for a dust (p. 84ff), if $\mathcal{U}_t$ is a compact region on V_t^3, its mass

$$\mathcal{M} = \int_{\mathcal{U}_t} \rho \omega_V^3$$

is constant under the flow. From Equation 12-31 we get

$$\mathrm{vol}''(t) = 12\pi\kappa\mathcal{M} - \int_{\mathcal{U}_t} R_{V_t^3} \omega_V^3. \tag{12-32}$$

In particular, *the volumetric expansion accelerates when* $R_{V_t^3} \leq 0$ (in fact for $R_{V^3} < 12\pi\kappa\rho$) *but decelerates for* $R_{V_t^3} > 12\pi\kappa\rho$. The reader may verify this behavior in the positively curved Friedmann models (Case II) by investigating the volume factor $G^3(t)$. (For further applications of the second and higher variations of volume see Frankel, 1975; and Brill and Flaherty, 1976.)

General Vorticity-Free Cosmologies: Singularities

Consider an incoherent dust, $p = 0$, with vorticity zero. Locally coordinates can be introduced so that the metric assumes the form of Equation 12-1 with t-lines as geodesic world lines of the fluid. By the generalized Poisson equation (Eqn. 12-24), with $g_{00} = -1$, the mean curvature H of the spatial sections satisfies

$$\partial H/\partial t = 4\pi\kappa(\rho + 3p) + \mathrm{tr}(\mathsf{b}^2) \geq \mathrm{tr}(\mathsf{b}^2). \tag{12-33}$$

$H = \mathrm{tr}(\mathsf{b}) = \kappa_1 + \kappa_2 + \kappa_3$ satisfies, via the Schwartz inequality

$$H^2 \leq 3(\kappa_1^2 + \kappa_2^2 + \kappa_3^2) = 3\mathrm{tr}(\mathsf{b}^2),$$

and so $\partial H/\partial t \geq H^2/3$, or

$$\frac{\partial}{\partial t} H^{-1} \leq -\frac{1}{3}. \tag{12-34}$$

Let $H(p_0, t)$ be the mean curvature of the spatial section at the event on the world line passing through p_0 at time $t = 0$. If $H(p_0, 0) < 0$, i.e., a neighborhood of p_0 is expanding volumetrically at $t = 0$, then H^{-1} will vanish at some time t_1, $|t_1| \leq 3(-H(p_0, 0))^{-1}$, in the past, along this world line, provided this world line can be extended that far into the past. Likewise, if $H(p_0, 0) > 0$, i.e., if there is volumetric contraction at $t = 0$, then H^{-1} will vanish at some time $t_2 < 3H^{-1}(p_0, 0)$ in the future.

In a dust model, from conservation of mass, (Eqn. 7-25)

$$u(\rho) + \rho \,\mathsf{div}\, u = d\rho/dt - \rho H = 0,$$

one can see rather easily, using $\partial H/\partial t \geq H^2/3$, that as $H \to \infty$, the density also $\to \infty$. Thus continuation of the world line through p_0 will lead to infinite densities (in the past, for expansion at $t = 0$, and in the future, for contraction at $t = 0$). These results are due, I believe, to Raychaudhuri.

There are thus singularities of some sort in the past or in the future. Either *physical* singularities, e.g., infinite densities, develop or else the manifold structure becomes incomplete (world lines in some sense are incomplete). Note also, from $\rho = \mathsf{T}(u, u) = (8\pi\kappa)^{-1}[\mathsf{Ric}(u, u) + \frac{1}{2}R]$ that infinities in *curvatures* develop as $\rho \to \infty$. This is far different from dust collapsing to infinite density in Minkowski space, with Minkowski space itself remaining flat.

In the case of a Friedmann universe, the Hubble parameter is, from Equations 12-6 and 12-9, $h = -\frac{1}{3}H$. For the case of our expanding universe, its "age" since the big bang is $\leq h^{-1}$ (in this highly idealized model *all* world lines exploded at the same time in the past). The best current estimates of h via the recession of the galaxies seems to give age $\leq h^{-1} \sim 18 \times 10^9$ years. For a slightly more refined analysis, keep the term $4\pi\kappa\rho$ in Equation 12-33; then $\partial H/\partial t \geq H^2/2 + \Delta/2$, where again $\Delta = 8\pi\kappa\rho - 3h^2$. In the case of a *positively* curved Friedmann universe

$\Delta > 0$ (see Eqn. 12-10), and consequently the age of this positively curved universe is $\leq \frac{2}{3}h^{-1} \sim 12 \times 10^9$ years.

The above results are for a dust, i.e., $p = 0$. Perhaps if we allowed pressure gradients, the pressure could prevent such singularities from developing. Also it might be that rotation (vorticity) could influence the occurrence of singularities. These possibilities are not encouraged by the theorems of Penrose and Hawking; under various not unreasonable assumptions on the nature of the space-time manifold, this manifold *must be incomplete in some sense, somewhere!* (See Hawking and Ellis, 1973, for details.)

Note that these conclusions rest on the analysis of the behavior of a family of time-like curves *via* the Raychaudhuri equations (Eqn. 12-15), and these equations are of a purely geometric nature. General relativity enters through the requirement $\mathsf{Ric}(u, u) = 4\pi\kappa(\rho + 3p) \geq 0$. It is believed that all normal matter satisfies the "strong energy condition" $\mathsf{Ric}(u, u) \geq 0$.

General Vorticity-Free Cosmologies: Closed Spatial Universes

Nothing in general relativity has intrigued the lay public more than Einstein's possibility of a closed, finite spatial universe. The Friedmann models with positive curvature, i.e., $\Delta = 8\pi\kappa\rho - 3h^2 > 0$, have spatial sections that, if complete, are covered by the 3-sphere, and so are compact. Contrary to what one sees in most popular books on relativity, there are also compact models with zero or negative constant curvature, e.g., the 3-torus in the case $\Delta = 0$. As was mentioned earlier, what *is* true is that these models, though locally isotropic, are not globally isotropic. For example, the negatively curved models are governed by the following (stated in this form by Nomizu).

Bochner's theorem In a Riemannian V^n with negative definite Ricci curvature, $\mathsf{Ric}(u, u) < 0$ for $u \neq 0$, no nonzero Killing vector X can achieve its maximum length. In particular, if V^n is compact, $X \equiv 0$.

We shall sketch a proof involving Jacobi fields, a concept that will be needed in our discussion of the geometry of a spatial section.

PROOF Suppose X achieves its maximum length at p. Let T be a unit vector at p and let $\gamma(s) = \exp_p(sT)$ be the geodesic thru T at p.

Let again T be the unit tangent to γ. Then $0 = T\langle X, X\rangle = 2\langle \nabla_T X, X\rangle$ at p, and further $T^2\langle X, X\rangle = 2\langle \nabla_T^2 X, X\rangle + 2\|\nabla_T X\|^2$. The Killing field X generates a one-parameter group $\{\varphi_\varepsilon\}$ of isometries, and so the varied curves $\varphi_\varepsilon\gamma$ are again geodesics. Let T be the tangents to the varied curves; we have, as in Chapter 7, $\nabla_T X = \nabla T/\partial\varepsilon$. A variation vector field X giving rise to a variation by geodesics is called a *Jacobi field*. From $\nabla_T^2 X = \nabla_T\nabla_T X = \nabla_T \nabla T/\partial\varepsilon = \mathsf{R}(T, X)T + \nabla/\partial\varepsilon(\nabla_T T) = \mathsf{R}(T, X)T$, we see that a Jacobi field satisfies the *Jacobi equation of geodesic deviation:*

$$\nabla_T^2 X + \mathsf{R}(X, T)T = 0. \tag{12-35}$$

Consequently,

$$\frac{1}{2}T^2\langle X, X\rangle = -\langle \mathsf{R}(X, T)T, X\rangle + \|\nabla_T X\|^2. \tag{12-36}$$

If we do this for $(n - 1)$ unit orthogonal vectors $T_1, \ldots, T_{n-1}$ at p, all orthogonal to X at p, we get there

$$\frac{1}{2}\sum_\alpha T_\alpha^2\langle X, X\rangle = -\|X\|^2\mathsf{Ric}(X, X) + \sum_\alpha \|\nabla_{T_\alpha} X\|^2 > 0$$

which contradicts $\|X\|$ having a maximum at p.

We now consider the problem of imposing natural conditions on a universe that will force the spatial sections to be compact. Let V_t^3 locally be a one-parameter family of hypersurfaces orthogonal to the flow vector $\partial/\partial t$ and let $X(t)$ be a vector field defined along the world line of the flow thru $p \in V_0^3$ that is invariant under the flow and orthogonal to it. We say that space is *expanding* (more accurately, noncontracting) at p if $d/d\tau\|X(t)\| \geq 0$ at $t = 0$ for all such X, and we say that the expansion is *decelerating* at p if $\frac{d^2}{d\tau^2}\|X(t)\| \leq 0$ at $t = 0$, for all such X. Here τ is proper time along the t-line.

Recall that in the case of a Friedmann model, i.e., constant spatial curvature, if $\Delta = 8\pi\kappa\rho - 3h^2 > 0$ then the spatial curvature is positive and consequently space is compact if it is complete (p. 149ff). Of course actual space is not exactly locally isotropic. It is remarkable that this condition $\Delta > 0$ in the locally isotropic case can be replaced by an only very slightly stronger condition $\Delta' > 0$ in the *general* case of a vorticity-free cosmological dust model.

Theorem (Galloway, 1976). Let M^4 be a cosmological model of a dust with vorticity zero. Let V_0^3 be a "maximal spatial section," i.e., a maximal connected hypersurface orthogonal to the velocity vector u. Assume

(1) V_0^3 is geodesically complete;

(2) Space is expanding at each $p \in V_0^3$;

(3) The expansion is decelerating at each $p \in V_0^3$;

(4) $\Delta' = \inf_{V_0^3}[4\pi\kappa\rho - 3h^2] > 0$,

where $h = ⅓ \operatorname{div} u = -H/3$ is the averaged Hubble expansion parameter at points of V_0^3. Then V_0^3 is compact with finite fundamental group and

$$\text{diameter}\,(V_0^3) \leq \pi \sqrt{\frac{2}{\Delta'}}\,\cdot$$

PROOF The proof is an application of the classic theorem of S. B. Myers that shows that positive Ricci curvature implies compactness (see, for example, Milnor, 1963, p. 105). We shall show that the Ricci curvature of V_0^3 satisfies $\mathsf{Ric}_{V_0} \geq \Delta'$. For this we need the following purely geometric result on space-like hypersurfaces in a pseudo-Riemannian M^4.

Lemma Let V_t^3 be a family of space-like hypersurfaces orthogonal to a geodesic congruence of curves. Let $e_\alpha(t)$, $\alpha = 1, 2, 3$, be invariant under the flow along the congruence (parameterized by proper time t) and such that $\{e_\alpha(0)\}$ form an orthonormal basis for the tangent space to V_0^3 at p. Then

$$\begin{aligned}\mathsf{Ric}_{V_0}(e_1, e_1) &= \mathsf{Ric}(e_1, e_1) + 2\mathsf{B}^2(e_1, e_2) + 2\mathsf{B}^2(e_1, e_3)\\ &\quad + \left[\frac{\partial\|e_1\|}{\partial t}\right]^2 - \left[\frac{\partial^2\|e_1\|}{\partial t^2}\right] + \left[\frac{\partial\|e_1\|}{\partial t}\right]\cdot H,\end{aligned}$$

where Ric is the Ricci tensor for M^4, B is the second fundamental form of V_0^3, and time derivatives are evaluated at $t = 0$.

PROOF Let $e_0 = \partial/\partial t = u$ be the unit tangent field to the geodesics, t being proper time starting with $t = 0$ on V_0^3. Then, at p, from Equation 4-5,

$$\mathsf{Ric}_V(e_1, e_1) = \mathsf{K}_V(e_1 \wedge e_2) + \mathsf{K}_V(e_1 \wedge e_3).$$

From the Gauss equations (Eqn. 4-11), we have

$$\mathsf{K}(e_\alpha \wedge e_\beta) = \mathsf{K}_V(e_\alpha \wedge e_\beta) - \varepsilon_0 \varepsilon_\alpha \varepsilon_\beta [\mathsf{B}(e_\alpha, e_\alpha)\mathsf{B}(e_\beta, e_\beta) - \mathsf{B}^2(e_\alpha, e_\beta)]$$

where, as usual,

$$\mathsf{B}(e_\alpha, e_\beta) = \langle \mathsf{b}(e_\alpha), e_\beta \rangle = \mathsf{B}(e_\beta, e_\alpha).$$

Thus

$$\begin{aligned} \mathsf{Ric}_V(e_1, e_1) = {} & \mathsf{K}(e_1 \wedge e_2) + \mathsf{K}(e_1 \wedge e_3) \\ & - \mathsf{B}(e_1, e_1)\mathsf{B}(e_2, e_2) + \mathsf{B}^2(e_1, e_2) \\ & - \mathsf{B}(e_1, e_1)\mathsf{B}(e_3, e_3) + \mathsf{B}^2(e_1, e_3), \end{aligned}$$

that is,

$$\begin{aligned} \mathsf{Ric}_V(e_1, e_1) = {} & \mathsf{Ric}(e_1, e_1) - \mathsf{K}(e_1 \wedge e_0) - \mathsf{B}(e_1, e_1)H \\ & + \mathsf{B}^2(e_1, e_1) + \mathsf{B}^2(e_1, e_2) + \mathsf{B}^2(e_1, e_3), \end{aligned} \tag{12-37}$$

since $H = \mathsf{B}(e_1, e_1) + \mathsf{B}(e_2, e_2) + \mathsf{B}(e_3, e_3)$ at p.

The curvature term $\mathsf{K}(e_1 \wedge e_0)$ can be evaluated as follows. Since e_0 is the unit tangent field to a geodesic flow, and e_1 is invariant under this flow, e_1 is a Jacobi field along the time lines and so satisfies the Jacobi equation (Eqn. 12-35).

$$\nabla_{e_0}\nabla_{e_0}e_1 = \mathsf{R}(e_0, e_1)e_0.$$

Thus, from Equations 4-3 and 12-36

$$\begin{aligned} \mathsf{K}(e_1 \wedge e_0) &= \varepsilon_1 \varepsilon_0 \langle \mathsf{R}(e_1, e_0)e_0, e_1 \rangle = -\langle \mathsf{R}(e_1, e_0)e_0, e_1 \rangle \\ &= \frac{1}{2}\frac{\partial^2}{\partial t^2} \langle e_1, e_1 \rangle - \|\nabla_{e_1}e_0\|^2 \\ &= \frac{1}{2}\frac{\partial^2}{\partial t^2} \|e_1\|^2 - \|\mathsf{b}(e_1)\|^2 \\ &= \left(\frac{\partial}{\partial t}\|e_1\|\right)^2 + \|e_1\| \frac{\partial^2 \|e_1\|}{\partial t^2} - \sum_\alpha \langle \mathsf{b}(e_1), e_\alpha \rangle^2 \\ &= \left(\frac{\partial}{\partial t}\|e_1\|\right)^2 + \frac{\partial^2 \|e_1\|}{\partial t^2} - \sum_\alpha \mathsf{B}^2(e_1, e_\alpha) \quad \text{at} \quad t = 0. \end{aligned}$$

Furthermore

$$\mathsf{B}(e_1, e_1) = \langle \mathsf{b}(e_1), e_1\rangle = -\langle \nabla_{e_0} e_1, e_1\rangle = -\frac{\partial}{\partial t}\|e_1\| \qquad (12\text{-}38)$$

at $t = 0$, and so

$$\mathsf{K}(e_1 \wedge e_0) = \frac{\partial^2\|e_1\|}{\partial t^2} - \mathsf{B}^2(e_1, e_2) - \mathsf{B}^2(e_1, e_3).$$

Putting this into Equation 12-37 yields the lemma.

Let us now introduce the notation

$$\|e_\alpha\|^{\boldsymbol{\cdot}} = \frac{\partial}{\partial t}\|e_\alpha\| \quad \text{at} \quad t = 0,$$

From Equation 12-38

$$H = -\|e_1\|^{\boldsymbol{\cdot}} - \|e_2\|^{\boldsymbol{\cdot}} - \|e_3\|^{\boldsymbol{\cdot}},$$

while

$$(\|e_1\|^{\boldsymbol{\cdot}})^2 + \|e_1\|^{\boldsymbol{\cdot}} H = -\|e_1\|^{\boldsymbol{\cdot}}\|e_2\|^{\boldsymbol{\cdot}} - \|e_1\|^{\boldsymbol{\cdot}}\|e_3\|^{\boldsymbol{\cdot}}.$$

Now using *expansion*,

$$\|e_1\|^{\boldsymbol{\cdot}}\|e_2\|^{\boldsymbol{\cdot}} + \|e_1\|^{\boldsymbol{\cdot}}\|e_3\|^{\boldsymbol{\cdot}} \le \sum_{\alpha<\beta}\|e_\alpha\|^{\boldsymbol{\cdot}}\|e_\beta\|^{\boldsymbol{\cdot}}$$

$$= \frac{1}{2}\left[\left(\sum_\alpha \|e_\alpha\|^{\boldsymbol{\cdot}}\right)^2 - \sum_\alpha (\|e_\alpha\|^{\boldsymbol{\cdot}})^2\right]$$

which by Schwarz's inequality is

$$\le \frac{1}{2}\left[\left(\sum_\alpha \|e_\alpha\|^{\boldsymbol{\cdot}}\right)^2 - \frac{1}{3}\left(\sum_\alpha \|e_\alpha\|^{\boldsymbol{\cdot}}\right)^2\right] \le \frac{1}{3}\left(\sum_\alpha \|e_\alpha\|^{\boldsymbol{\cdot}}\right)^2.$$

Using $\left(\sum_\alpha \|e_\alpha\|^{\boldsymbol{\cdot}}\right)^2 = H^2 = 9h^2$ we have then from the lemma

$$\mathsf{Ric}_{V_0}(e_1, e_1) \ge \mathsf{Ric}(e_1, e_1) - \frac{1}{3}\cdot 9h^2 - \frac{\partial^2}{\partial t^2}\|e_1\|$$

and *deceleration* then yields

$$\text{Ric}_{V_0}(e_1, e_1) \geq \text{Ric}(e_1, e_1) - 3h^2. \tag{12-39}$$

For a dust, relative to the basis $e_0 = u, e_1, e_2, e_3$

$$T^{ij} = \rho\, u^i u^j,$$

and so $\mathsf{T}(e_\alpha, e_\alpha) = 0$, $\mathsf{T}(e_0, e_0) = \rho$, $T = -\rho$. Thus

$$\text{Ric}(e_1, e_1) = 8\pi\kappa \left[\mathsf{T}(e_1, e_1) - \frac{1}{2}T\right] = 4\pi\kappa\rho$$

and consequently

$$\text{Ric}_{V_0}(e_1, e_1) \geq 4\pi\kappa\rho - 3h^2$$

as desired.

We shall conclude with three final remarks.

First, note that the original Einstein-Friedmann models were assumed of the form $R \times V^3$, i.e., it was assumed that there is a *global* time coordinate t such that the spatial sections V_t^3 orthogonal to the time lines are obtained by putting t = constant. No such assumption is made in Galloway's theorem; the spatial section V_0^3 is merely a maximal leaf of the foliation orthogonal to the time lines and this maximal leaf could possibly intersect the same time line many times. Recall that we expect singularities to develop along time lines. Let us say, then, that *time* ε *exists for* V_0^3 if each time line that starts on V_0^3 can be continued for all (proper) time $|t| \leq \varepsilon$. Since Galloway's V_0^3 is compact, time ε exists for V_0^3 for some $\varepsilon > 0$.

Corollary Let M^4 be time-orientable. If time ε exists for V_0^3, then no time line $t \to \gamma(t)$ with $\gamma(0) = p \in V_0^3$ meets V_0^3 again for $0 < |t| \leq \varepsilon$.

PROOF We shall use the fact that V_0^3 is a compact leaf for the foliation orthogonal to the (geodesic) time lines of the vorticity-free dust of density $\rho > 0$. Assume then for some $p \in V_0^3$ that $\gamma(t') \in V_0^3$ for some $t' \neq 0$, $|t'| \leq \varepsilon$. Translate V_0^3 in time by amount t'. Since V_0^3 is a leaf orthogonal to the geodesic time lines, its translate is, by

Gauss's lemma, again a leaf. Since $\gamma(t') \in V_0^3$, this leaf is again V_0^3. This means that time $2|t'|$ exists for V_0^3 and translating V_0^3 by $2t'$ gives still another point on V_0^3 lying on the time line γ. Continuing in this fashion we see that time $n|t'|$ exists for V_0^3 for all positive integers n. This contradicts Raychaudhuri's result (p. 158) since either $H(p) \neq 0$ or Equation 12-33 will show that $H(\gamma(t)) \neq 0$ for small t.

Second, we point out that Galloway has obtained a generalization of his closure theorem which allows nongeodesic motions (in particular, fluids with pressure gradients would be allowed). In Galloway's theorem leave assumptions (1), (2), and (3) as they are. Condition (4) is replaced by

$$\Delta'' = \inf_{V_0^3}[\mathrm{Ric}(X, X) - 3h^2] > 0$$

for all unit X tangent to V_0^3. (Note that Ric is the Ricci tensor for M^4, and in the case of a perfect fluid $\mathrm{Ric}(X, X) = 4\pi\kappa(\rho - p)$.) There is a final condition

$$\sup_{V_0^3}\| \nabla_u u \| = a < \infty$$

that the flow lines have bounded acceleration on V_0^3. Galloway then concludes that V_0^3 is compact with

$$\text{diameter } (V_0^3) \leq \pi \cdot \frac{a + \sqrt{a^2 + 2\Delta''}}{\Delta''}.$$

The proof, in the method of S. B. Myers, is to show that a geodesic on V_0^3 (in the induced Riemannian metric) whose length is greater than the right side of the above diameter formula cannot be minimizing. The metric of M^4 near V_0^3 is locally of the form

$$ds^2 = g_{00}\, dt^2 + dl^2$$

$$dl^2 = g_{\alpha\beta}\, dx^\alpha\, dx^\beta, \text{ the metric on } V_0^3.$$

Let γ be a dl^2 geodesic on V_0^3, with unit tangent e_1. A computation similar to that for the lemma in Galloway's theorem yields a formula for $\mathrm{Ric}_V(e_1, e_1)$ of a more complicated nature. Using the expansion and deceleration assumptions, Galloway finds that along γ

$$\mathrm{Ric}_{V_0}(e_1, e_1) \geq \mathrm{Ric}(e_1, e_1) - 3h^2 + \frac{d^2}{dl^2} \log \sqrt{-g_{00}(\gamma(l))}. \quad (12\text{-}40)$$

Galloway then derives the following refinement of Myers's principal lemma.

> **Lemma** (Galloway) Let γ be a geodesic segment joining two points in a Riemannian V^n. Let l be arc length and let T be the unit tangent along γ. Suppose that there is a continuously differentiable function f defined along γ, $|f| \leq a$ on γ, and a constant $\delta > 0$ such that $\mathrm{Ric}(T, T) \geq \delta + df/dl$ along γ. Then γ cannot be minimizing if the length of γ is greater than
>
> $$\pi \frac{a + \sqrt{a^2 + (n-1)\delta}}{\delta}.$$

It is important that $\delta + df/dl$ need not be positive. This lemma is then applied, using Equation 12-40, with $\delta = \Delta''$ and $f(l) = d/dl \log \sqrt{-g_{00}(\gamma(l))}$. Along γ

$$|f| \leq \|\mathrm{grad}_V \log \sqrt{-g_{00}}\| = \|\nabla_u u\|$$

by the lemma on page 129. (For details, see Galloway, 1979.)

Finally, note that compactness of the spatial section can be proved with a weakened form of either the original or revised condition (4), namely,

$$\lim_{R\to\infty} \int_0^R [\mathrm{Ric}(X, X) - 3h^2]dl = +\infty$$

for each V_0^3-geodesic emanating from a given point of V_0^3. This follows from the analysis (Eqn. 12-40) together with a theorem of Ambrose (1957). This weakened condition, however, does not allow one to estimate the diameter of the space. [A similar analysis can be applied to universes that are static or "rotating" but not expanding (Frankel, preprint).] Compactness and finiteness of the fundamental group further imply, via the Reeb stability theorem (see G. Reeb, 1952) that all leaves V_t^3 near V_0^3 are diffeomorphic to V_0^3. One can also show by foliation methods that the "achronality" of V_0^3, as expressed in the corollary of p. 164, again holds provided we assume a "causality" condition, that is, provided we assume that there are no closed timelike lines in the universe.

References

Adler, R., M. Bazin, and M. Schiffer. 1975. *Introduction to General Relativity*, 2nd edition. New York: McGraw-Hill.

Ambrose, W. 1957. A theorem of Myers. *Duke Journal,* **24:**345–348.

Bergmann, P. G. 1942. *Introduction to the Theory of Relativity*. Englewood Cliffs, N.J.: Prentice-Hall.

Brill, D., and F. Flaherty. 1976. Isolated maximal surfaces in spacetime. *Com. Math. Phys.,* **50:**157.

de Rham, G. 1955. *Variétés Différentiable*. Herman.

Ehlers, J., F. Pirani, and A. Schild. 1972. The geometry of free-fall and light propagation. In *General Relativity,* L. O'Raifeartaigh, ed. Oxford University Press.

Einstein, A. 1931. *Relativity: The Special and General Theory*. New York: Crown.

———. 1950. *The Meaning of Relativity*. Princeton University Press.

Frankel, T. 1975. Applications of Duschek's Formula to cosmology and minimal surfaces. *Bull. Amer. Math. Soc.,* **81:**579–582.

———. Preprint. On the size of a rotating fluid mass in general relativity.

Galloway, G. 1976. *Studies in Gravitation,* thesis, University of California, San Diego.

———. 1979. A generalization of Myers' Theorem and an application to relativistic cosmology. *J. Diff. Geom.,* in press.

Gamow, G. 1947. *Mr. Tompkins in Wonderland.* New York: Macmillan.

Hawking, S., and G. Ellis. 1973. *The Large Scale Structure of Space-Time.* Cambridge University Press.

Lanczos, C. 1974. *The Einstein Decade.* New York: Academic Press.

Lovelock, D. 1972. The four-dimensionality of space and the Einstein tensor. *J. Math. Phys.,* **13**(6):874–876.

MacCallum, M. 1971. A class of homogeneous cosmological models III. *Com. Math. Phys.,* **20**:58.

Mehra, J. 1974. *Einstein, Hilbert and the Theory of Gravitation.* Hingham, Mass.: Reidel.

Milnor, J. 1963. *Morse Theory.* Princeton University Press.

Misner, C., K. Thorne, and J. Wheeler. 1973. *Gravitation,* San Francisco: W. H. Freeman and Company.

Reeb, G. 1952. Sur certain proprietées topologíques de variétés feuilletées. *Actual Sci. Ind.* No. 1183, Hermann, Paris.

Synge, J. 1960. *Relativity: The General Theory.* Amsterdam: North Holland.

Tolman, R. 1934. *Relativity, Thermodynamics, and Cosmology.* Oxford University Press.

Warner, F. 1971. *Foundations of Differentiable Manifolds and Lie Groups.* Glenview, Ill.: Scott Foresman.

Weyl, H. 1952. *Space-Time-Matter.* New York: Dover.

Index